Sustainable Development Teaching

The aim of this book is to support and inspire teachers to contribute to much-needed processes of sustainable development and to develop teaching practices and professional identities that allow them to cope with the specificity of sustainability issues and, in particular, with the teaching challenges related to the ethical and political dimension of environmental and sustainability education.

Bringing together recent scholarship on the topic, this book translates state-of-the-art academic research into teaching models, methods and tools. Starting with an outline of the challenge of sustainability, it offers insights and models for understanding the interesting yet ambiguous concept of 'sustainable development' and the complex process of transforming society in a more sustainable direction (Part I). It then goes on to provide a guide to preparing courses and lessons as well as tools for reflection about teaching practices and the multiplicity of approaches to addressing ethical and political challenges in sustainable development teaching (Part II). Finally, the book offers useful conceptual frameworks, models and typologies about the concrete design and implementation of sustainable development teaching (Part III).

This book will be essential reading for students of education, as well as teachers in compulsory and higher education and sustainability education researchers.

Katrien Van Poeck is a senior researcher on environmental and sustainability education at Ghent University's Centre for Sustainable Development, Belgium.

Leif Östman is Professor of Curriculum Studies at Uppsala University's Department of Education, Sweden.

Johan Öhman is Professor of Education at Örebro University's School of Humanities, Education and Social Sciences, Sweden.

Routledge Studies in Sustainability

For more information about this series, please visit: www.routledge.com/Routledge-Studies-in-Sustainability/book-series/RSSTY

Sustainable Development Teaching

Ethical and Political Challenges

Edited by Katrien Van Poeck,
Leif Östman and Johan Öhman

LONDON AND NEW YORK

First published 2019
by Routledge
2 Park Square, Milton Park, Abingdon, Oxon OX14 4RN

and by Routledge
605 Third Avenue, New York, NY 10017

First issued in paperback 2020

Routledge is an imprint of the Taylor & Francis Group, an informa business

British Library Cataloguing-in-Publication Data
A catalogue record for this book is available from the British Library

Library of Congress Cataloging-in-Publication Data
Names: Poeck, Katrien Van, editor. | Ostman, Leif, editor. | Ohman,
    Johan, editor.
Title: Sustainable development teaching : ethical and political challenges /
    edited by Katrien Van Poeck, Leif Ostman and Johan Ohman.
Description: Abingdon, Oxon ; New York, NY : Routledge, 2019. |
    Series: Routledge studies in sustainability | Includes bibliographical
    references and index.
Identifiers: LCCN 2019002674 (print) | LCCN 2019014454 (ebook) |
    ISBN 9781351124348 (eBook) | ISBN 9780815357537 (hbk) |
    ISBN 9781351124348 (ebk)
Subjects: LCSH: Sustainability—Study and teaching. | Sustainable
    development—Study and teaching. | Environmental education—
    Moral and ethical aspects. | Environmental education—Political
    aspects.
Classification: LCC GE196 (ebook) | LCC GE196 .S874 2019 (print) |
    DDC 304.2071—dc23
LC record available at https://lccn.loc.gov/2019002674

ISBN 13: 978-0-367-72958-5 (pbk)
ISBN 13: 978-0-8153-5753-7 (hbk)
ISBN 13: 978-1-351-12434-8 (ebk)

Typeset in Goudy
by Apex CoVantage, LLC

Contents

Figures

Tables

Textboxes

Preface

It has not been difficult to find the motivation for developing a book that aims to encourage teachers and future teachers to take on a deeper sense of commitment towards environmental and sustainability education (ESE). Sustainable development is with no doubt one of the most important tasks of our time and education certainly plays a key role here. But sustainable development is also an educational challenge: how do we educate about problems without clear-cut solutions that often involve ethical and political controversy? We are fortunate to have a number of skilled colleagues who have devoted their work to this challenge, in both their research and teaching, and have agreed to share their experiences in this book. In this way, the book can be seen as the result of collaborative research in and development of ESE for more than two decades.

ESE has always been one of the key research themes of SMED – Studies of Meaning-making in Educational Discourses, a cross-university research environment founded in 2003 by Leif Östman and Johan Öhman and five other colleagues. SMED consists of over 50 researchers and doctoral students at Uppsala University and Örebro University but also collaborates closely with scholars from other universities. The editors and many of the contributors of this book are members of this research environment. One of SMED's appreciated contributions to the field of didactics and educational science has been the development of analytical models and methods and its application in numerous empirical studies. This theoretical and empirical basis allowed us to develop this book.

Besides SMED, other scientific collaborations and networks have also greatly contributed to the creation of this book. Many of the contributors have been involved – as a leader/supervisor or as a (doctoral) researcher – in the Swedish Graduate School in Education and Sustainable Development managed by Uppsala University (GRESD, 2009–2016). GRESD aimed to enhance knowledge production about the complexity of teaching and learning about sustainable development, including the challenges involved in dealing with its ethical and political dimensions. Empirical research, rather than ideological debate, has always been a major concern in GRESD. Since 2016, the editors and several of the authors are collaborating in the international research networks 'SEDwise – Sustainability Education: Teaching and learning in the face of wicked socio-ecological problems' and 'Public pedagogy and sustainability challenges' coordinated by Ghent

University's Centre for Sustainable Development. In both networks, the ethical and political dimensions of ESE take central stage and the teaching challenges this brings about are addressed from a theoretical as well as practical perspective. The ESER (environmental and sustainability education research) network of the European Educational Research Association is another academic network in which the authors of this book are actively engaged. Its overall objective is to share, discuss, disseminate and advance ESE research. Besides countless research projects and articles, an important source of inspiration for the book has been the collaboration created in developmental projects around the world as well as over ten years of cooperation within international training programmes on environmental and sustainability teaching.

All these collaborations and networks have made it possible to present and discuss the research and ideas elaborated in this book at seminars and conferences and to engage in scientific dialogues with a number of great scholars around the world who we have many reasons to thank for their inspiration, deep theoretical insights and seminal educational points of view: Joacim Andersson, Erik Andersson, Stefan Bengtsson, Lovisa Bergdahl, Luiz Marcelo de Carvalho, Jim Garrison, Jeppe Læssøe, Elsa Lee, Heila Lotz-Sisitka, Elisabet Langmann, Jonas Andreasen Lysgaard, Greg Mannion, Jan Masschelein, Marcia McKenzie, Jutta Nikel, Rob O'Donoghue, Phillip Payne, Alan Reid, Carl Anders Säfström, Bill Scott, Chris Shilling Maarten Simons, Per Sund, Tomasz Szkudlarek, Joke Vandenabeele, Arjen Wals and Danny Wildemeersch.

It is our sincere hope that this book will stimulate teacher discussions on the interface between ethics, politics and education and also encourage the continued progress of educational practices in the field of sustainable development.

Katrien Van Poeck, Leif Östman and Johan Öhman

Gent, Uppsala and Örebro, January 2019

Contributors

Andersson, Pernilla – Department of Humanities and Social Sciences Education, Stockholm University

Pernilla Andersson is senior lecturer in social sciences education at the Department of Humanities and Social Sciences Education at Stockholm University. She teaches subject matter didactics in teacher education, at the master's programme in social sciences education and in third cycle studies. She did her doctoral studies in environmental studies at Södertörn University, studying the integration of 'sustainable development' in business education. Her research interests involve sustainability education, pluralist approaches to economics education and the development of research methods for studies in and of different subject specific teaching and learning in classroom practices.

Block, Thomas – Centre for Sustainable Development, Ghent University

Thomas Block is director of the Centre for Sustainable Development and professor of 'Sustainability and Governance' at the Department of Political Sciences (Ghent University, Belgium). His research approach involves the use of a (nuanced) constructivist epistemology, a complexity-acknowledging perspective, an interpretative policy analysis framework, a participative research design and the framing of sustainability issues as a 'political' matter. His research focus is on complex decision-making and transition governance, education on wicked sustainability issues, scenarios and future studies, sustainable cities and urban projects. He is lecturer of courses such as 'Politics of Sustainability', 'Sustainable Cities' and the university-wide elective course 'Sustainability Thinking'.

Håkansson, Michael – Department of Education, Stockholm University

Michael Håkansson is working as a senior lecturer in Didactics at the Department of Education, Stockholm University, Sweden. His research interest is how conflictual social relations can be used in environmental and sustainability education to develop as sustainable and fair social relations as possible. Before becoming a researcher in education, he worked for ten years as a teacher in social science at a Swedish lower secondary school.

Hansson, Petra – Department of Teacher Education and School Research, Oslo University

Petra Hansson has a PhD in education from Uppsala University. She is currently holding a Post Doc at the Department of Education and School Research at the University of Oslo, Norway. Hansson's research interest is Environmental and Sustainability Education with a focus on aesthetic experiences and meaning-making. In particular, her research focuses on student encounters with literary texts. Lately, she has been engaged in research about public pedagogy and sustainable development in museum settings. In her research, Hansson has combined empirical discourse analyses with conceptual and methodological development.

Kronlid, David O. – Department of Education, Uppsala University

David O. Kronlid is associate professor in Ethics and a senior lecturer in Education at the Department of Education, Uppsala University. He is the research leader at the Swedish International Centre of Education and Sustainable Development (SWEDESD). He has conducted interdisciplinary ethics research around ecofeminism, mobility, climate change justice and education for sustainable development (ESD). In collaboration with national and international partners, he has explored theorising and methods around cross-disciplinary sustainable development higher education and research, developed ESD teaching and learning models for value education, and innovated problem-solving process tools for municipalities and ESD-agents in post-normal situations.

Lundegård, Iann – Department of Mathematics and Science Education, Stockholm University

Iann Lundegård is an associate professor in Science Education at the Department of Mathematics and Science Education, Stockholm University. His research interest is educational philosophy associated with students' deliberations and meaning-making on sustainable development. He has a long experience in working with pre-service and in-service teacher training in science education and has written several textbooks and contributed to curriculum development at the Swedish Agency of Education.

Mårdh, Andreas – School of Humanities, Education and Social Sciences, Örebro University

Andreas Mårdh is a PhD candidate at Örebro University and a member of the research environment SMED (Studies of Meaning-making in Educational Discourses). His main research interests are in the political dimension of history curriculum and the role that emotions play in citizenship education.

Öhman, Johan – School of Humanities, Education and Social Sciences, Örebro University

Johan Öhman is professor of education at Örebro University's School of Humanities, Education and Social Sciences and is one of the founders of the research group SMED (Studies of Meaning-making in Educational Discourses). His work is based on John Dewey's pragmatic philosophy and view of the democratic potential of education. His area of research is ethical and democratic issues within the sphere of sustainability and environmental education, especially learning outcomes of student discussions, students' argumentation and teacher-student interactions. He has a background as a biology teacher in secondary school and has been involved in teacher training since 1993.

Öhman, Marie – Department of Health Sciences, Örebro University

Marie Öhman has a PhD in Sociology and is professor in Sport Science with a focus on didactics/education at Department of Health Sciences, Örebro University, Sweden. She has worked as a teacher educator since 1993 and her main areas of interest are the everyday lives of teachers and students in school. In her research, classroom observations and how teaching relates to norms, values and sustainability issues are prominent. Drawing on Michel Foucault's work, several of her publications have also focused on 'physical touch', health, body, power relations, governing processes and socialisation of the 'good' citizen.

Östman, Leif – Department of Education, Uppsala University

Leif Östman is professor of curriculum studies at Uppsala University. He is the director and one of the founders of the research group SMED (Studies of Meaning-making in Educational Discourses). In his research in science education and environmental and sustainability education, he has developed theoretical models and analytical methods based on pragmatist philosophy and the later works of Ludwig Wittgenstein. He has a background as a science teacher in a lower secondary school. Since 1987 he has been involved in teacher training as well as different forms of cooperation with teachers, schools, NGOs and governmental departments in Sweden and internationally.

Paredis, Erik – Centre for Sustainable Development, Ghent University

Erik Paredis is an assistant professor at the Department of Political Sciences of Ghent University, where he is connected to the Centre for Sustainable Development and the Ghent Institute for International Studies. His research interests include the politics of sustainability, the governance of transitions, the role of civil society and the relationship between sustainability and science and technology. Empirically, his research has addressed waste and the circular economy, the agro-food system, and the housing and building system. He teaches courses such as 'Politics of Sustainability' and 'Sustainability Thinking'.

Pashby, Karen – School of Childhood, Youth and Education Studies, Manchester Metropolitan University

Karen Pashby is reader of Education Studies at Manchester Metropolitan University. She teaches undergraduates and postgraduates in the School of

Childhood, Youth and Education Studies and is a core member of the Education and Social Research Institute. She is adjunct professor in the Department of Educational Policy Studies at University of Alberta and docent in the Faculty of Educational Sciences at University of Helsinki. A former secondary educator (in Canada and Brazil) and experienced teacher educator, her research draws on postcolonial and decolonial theoretical resources to examine productive pedagogical tensions in education for global citizenship in multicultural contexts.

Rudsberg, Karin – Department of Humanities, Education and Social sciences, Örebro University

Karin Rudsberg is a senior lecturer at the Department of Humanities, Education and Social Sciences, Örebro University, Sweden. She has a PhD in curriculum studies from Uppsala University and her thesis is entitled 'Students' learning in argumentative discussions about sustainable development'. She is also a member of the research group SMED (Studies of Meaning-making in Educational Discourses). Her research focuses on learning processes, social interplay and the role of students' prior knowledge in education relating to sustainability- and socioscientific issues and mathematics education.

Sund, Louise – School of Humanities, Education and Social Sciences, Örebro University

Louise Sund is an experienced secondary school teacher, a teacher educator at Mälardalen University's School of Education, Culture and Communication, and a researcher in Education at Örebro University's School of Humanities. She is a member of SMED (Studies of Meaning-making in Educational Discourses), a cross-university research group in the field of didactics and educational science at Örebro University. She has an interest in environmental and sustainability education and citizenship education. Her research interests include philosophical perspectives and approaches to education and sustainable development.

Tryggvason, Ásgeir – School of Humanities, Education and Social Sciences, Örebro University

Ásgeir Tryggvason has a PhD in education and is an assistant lecturer at Örebro University, Sweden. His research interest is in the political dimension of education. With inspiration from educational philosophy and political theory, he explores the role of emotions, identities and conflicts in education. He has published articles about agonism and populism as approaches to the political dimension of education and is currently working with the question of hope in environmental and sustainability education.

Van Poeck, Katrien – Centre for Sustainable Development, Ghent University

Katrien Van Poeck is a senior researcher at Ghent University where she coordinates the Centre for Sustainable Development's research track on sustainability education. She conducts and supervises research on experiential learning in the context of sustainability transitions and on sustainability in higher education. Theoretically, she draws on scholarship in educational theory, political theory and science and technology studies and focuses at the interrelatedness of educational and political processes and challenges. Methodologically and empirically, pragmatist analytical methods largely inspire her work. She develops in-service training for sustainability education practitioners and teaches in a university-wide elective course on 'Sustainability Thinking'.

Acknowledgements

We would first and foremost like to thank our co-authors for their tremendous efforts and for the valuable contributions their work has resulted in.

Many thanks to all the teachers and students who have provided us access to their classrooms and made the empirical studies which lay the basis of this book possible.

Furthermore, we owe great thanks to the anonymous reviewers (teachers and teacher trainers) who provided valuable feedback on earlier versions of the chapters of this book.

We would also like to thank some of the most important financing sources which have provided the resources necessary for our research work: The Swedish Research Council, the Swedish National Agency of Education, Formas and the Research Foundation Flanders. The funding of Ghent University for the international research network SEDwise (Sustainability Education: Teaching and learning in the face of wicked socio-ecological problems) allowed us to organise editors' meetings and seminars with the authors during the process of developing this book.

Thanks, too, to all the members of our different networks who have faithfully attended seminars and conference sessions and shared their valuable viewpoints and comments on our research papers and drafts of the chapters of this book.

Introduction

Sustainable development teaching – ethical and political challenges

Katrien Van Poeck, Leif Östman and Johan Öhman

A book on sustainable development teaching: aims and focus

Over the last decades, we have increasingly been confronted with extensive and often complex sustainability issues: climate change, food crises, poverty, urbanisation, inequality, loss of biodiversity, migration, etc. Worldwide, a consensus grows that the pursuit of sustainable development is one of the major societal challenges of our times and that education plays a vital role in tackling it. In 2015, the United Nations adopted the 2030 Agenda for Sustainable Development: 17 sustainable development goals (SDGs) to transform our world by ending poverty, protecting the planet and ensuring prosperity for all. An utmost challenging ambition! Since it proves to be very difficult to come up with adequate solutions for sustainability problems, realising a more sustainable world is often considered to be a matter of learning to find a way out of unsustainability (Van Poeck et al. 2018). Or, as UNESCO (2014a) phrases it, 'sustainable development begins with education'. Hence, one of the SDGs is to realise quality education for all and to ensure that all students acquire knowledge and skills needed to promote sustainability. Moreover, education is considered a vital catalyst for all the other SDGs. If teaching and learning are, indeed, indispensable elements for achieving social justice and ecological sustainability, teachers throughout the world merit the best possible support.

The specificity of sustainability problems challenges traditional educational practices. Expert knowledge concerning these issues sometimes fails to offer us an unambiguous and uncontested foundation for decision-making (Ashley 2000). Furthermore, social and political controversy arises due to a lack of agreement on norms and values at stake and on the acceptability of proposed solutions (Lundegård and Wickman 2007; Knutsson 2013; Sund and Öhman 2014; Håkansson et al. 2017). As such, sustainability issues are increasingly characterised as problems without clear-cut solutions, which is at odds with common conceptions of teaching and learning in terms of transferring reliable knowledge and acquiring specific skills, values and attitudes. In such a context, it becomes vital to approach teaching and learning from a pluralistic perspective where students can critically assess a number of different ideas and ways of handling sustainability problems (Öhman 2008; Östman 2010). Acknowledging this, however, should not be conflated with an 'anything-goes' approach. The far-reaching consequences of sustainability problems underline the importance of not falling into undue relativism – a challenge which is particularly relevant in our times that have recently been labelled and criticised as a 'post-truth' era (Van Poeck 2018). When, on the one hand, single 'right' answers do not exist and, on the other, the seriousness and urgency of the problems faced cannot be neglected, questions arise about what and how to teach. Teachers are called to take on new challenges. Sustainable development teaching is thus a matter of finding and implementing appropriate ways to deal with knowledge, (un) certainty, values and norms, ethical dilemmas, political controversies, concerns for the planet and its inhabitants, etc. Obviously, this is not only a cognitive but also an ethical and political challenge involving values and feelings. As we will elaborate throughout the book, ethical challenges in sustainable development teaching have to do with how to handle the ethical dimension of sustainability issues, i.e. the good values that people find desirable and the right actions that reflect these values (see chapters 6 and 7). Political challenges, then, involve the political dimension of sustainable development, i.e. the question of how to organise society in a sustainable way, acknowledging that this inevitably requires judgements, prioritisations and decision-making about different and competing alternatives (see chapters 8 and 9).

The aim of this book is to support and inspire teachers to contribute to much-needed processes of sustainable development, to develop teaching practices and professional identities that allow them to cope with the specificity of sustainability issues and, in particular, with the teaching challenges related to the ethical and political dimension of environmental and sustainability education (ESE). Therefore, we have brought together recent scholarship on these questions. The authors who contributed to this edited collection are all active researchers in this field, but a majority of them are also teachers with a long teaching experience at different educational levels and school forms. They all share an understanding of sustainable development teaching as a practice that involves not only cognitive but also ethical and political challenges. By translating state-of-the-art academic research into teaching models, methods and tools, it is our hope that this book can be a support for handling ethical and political challenges in ESE teaching, nourish reflection, inspire innovative teaching practice and offer a coherent theoretical framework and a wide range of empirically grounded knowledge that is useful for teacher training programmes and ESE courses.

In the next sections, we will subsequently explain how this book is underpinned by educational research, describe how its development has taken shape with a teacher perspective in the forefront and give an overview of the content of the different chapters.

Background: pragmatist educational research

The book's central theme – the ethical and the political dimension of ESE and the teaching challenges this brings about – has long been a common research interest of the authors and editors. All of them have been engaged in research environments and networks with this specific focus. The topic has always been one of the key research themes of SMED – Studies of Meaning-making in Educational Discourses, a cross-university research environment founded in 2003. Much of the research within this environment has been inspired by pragmatist philosophy and especially John Dewey's concept of transaction. SMED has contributed to the international research field of ESE research through the development of analytical models and methods by applying transactional pragmatist theory to the domain of didactic research and through the application of these models and methods in numerous empirical studies. The substantial body of research that has thus been created served as the theoretical and empirical underpinning of this book and allowed us to develop a transactional theory of teaching and learning (elaborated in detail in chapters 3, 10 and 11). But, as mentioned in the preface, there are many researchers worldwide that have been a source of inspiration for the research that is presented in this book.

Foreground: a teacher perspective

As argued, this book's major ambition is to translate state-of-the-art academic research into practical knowledge with the aim to support teachers to handle ethical and political challenges in ESE practice. Hence, a teacher perspective was consistently at the forefront in its development. The past decades were marked by the birth of several ESE research journals (e.g. *Environmental Education Research, The Journal of Environmental Education, Journal of Teacher Education for Sustainability, Journal of Education for Sustainable Development, International Journal of Sustainability in Higher Education*) and a growing number of academic and professionally oriented books. However, very few scholarly contributions focus on teachers and teaching practice. This is surprising, as it can hardly be ignored that high-quality teaching is a vital condition for education to successfully realise its purposes, e.g. to contribute to a more sustainable world.

Sustainable development has become a topic that every teacher needs to take into consideration in one way or another. A number of influential policy initiatives has put ESE prominently on the international agenda. In 2002, the United Nations General Assembly proclaimed the UN Decade of Education for Sustainable Development (DESD, 2005–2014), emphasising that 'education is an indispensable element for achieving sustainable development'. At the end of this decade, UNESCO's Global Action Programme (GAP) on education for sustainable development has been launched as a follow-up to the DESD with the aim 'to generate and scale up action in all levels and areas of education and learning to accelerate progress towards sustainable development' (UNESCO 2014b). Both the DESD and the GAP highlight the crucial role of teachers in reorienting education toward sustainability. Priority Action Area 3 of the GAP, for instance, is focused on increasing the capacities of educators and trainers to more effectively deliver sustainability education. Also, the UN Sustainable Development Goals (SDGs) highlight the importance of sustainability education. Targets by 2030 to realise SDG 4 – quality education – are, for instance, to ensure that all learners acquire the knowledge and skills needed to promote sustainable development and to substantially increase the supply of qualified teachers. Increasingly, these international policy initiatives trickle down into national curricula, learning objectives, syllabuses, textbooks, courses, etc. In schools worldwide, sustainable development is a substantial part of the subject matter of several school subjects and an important topic for cross-curricular projects. Hence, the challenges involved in ESE are in different ways a concern for every teacher. To provide a theoretical basis and facilitate creativity in handling these challenges, we suggest that this book can be used for different kinds of training courses and programmes such as pre- and in-service teacher training, higher education faculty training, ESE courses and programmes and professional development programmes for educators, trainers and staff members of various ESE organisations.

As shown in the author biographies in the list of contributors, the editors and authors of this book have extensive experience in teaching, be it in compulsory schools, higher education and/or pre- and in-service teacher training. While writing our contributions, we have thus simultaneously drawn on the theoretical and empirical outcomes of our scientific research and on our practical experience as teachers and participants in numerous developmental projects around the world.

A brief overview of the content

The book consists of three parts, each with a specific focus and purpose.

Part I: education and the challenge of building a more sustainable world

Part I of this book broadly outlines the challenge of building a more sustainable world and raises some questions and concerns regarding how to understand and develop adequate educational practices in this respect. It offers an overview of insights and knowledge that allows for an elaborated understanding of the concept 'sustainable development', of the complex process of transforming society in a more sustainable direction and of the specificity of sustainability issues that are often characterised as 'wicked problems'. Furthermore, it addresses the question what those insights imply in relation to education, i.e. when it comes to teaching and learning about such complex processes and issues.

In chapter 1, Thomas Block and Erik Paredis discuss four misunderstandings about sustainability and transitions. These popular and often-used concepts are subject to a diversity of interpretations and misunderstandings about their meaning. Four common 'misunderstandings' are that sustainability is only about ecological concerns, that we need a waterproof and objective definition of sustainability, that every change towards sustainability can be seen as a transition and that we can easily plan and manage sustainability transitions. The authors argue that sustainability is neither an objective standard nor an arbitrary concept, but that it is a broad political notion. Through a brief introduction to the multilevel perspective on sustainability transitions, they also indicate that building a more sustainable world is a very complex process that requires fundamental changes in socio-technical regimes. Transitions, they argue, go hand in hand with resistance, power struggles and questions of legitimacy which constitute its political dimension.

Chapter 2 is written by Thomas Block, Katrien Van Poeck and Leif Östman. It elaborates how, due to some specific features that characterise many sustainability problems, sustainable development teaching does not only involve questions of knowledge but also ethical and political challenges. Starting from concrete examples and a typology of sustainability problems, the authors describe how factual uncertainty (scientific controversy) and disagreement on values and norms (ethical and political controversy) bring about specific challenges for knowledge production about sustainability problems and possible solutions. Seeking inspiration in proposals for new approaches to science such as 'post-normal science', they articulate the ethical and political challenges involved in sustainable development teaching and explore what it could mean to design 'post-normal education' for dealing with wicked sustainability problems in which facts are uncertain, values in dispute, stakes high and decisions urgent.

In chapter 3, editors Leif Östman, Katrien Van Poeck and Johan Öhman go deeper into the challenges for teaching and learning brought about by sustainability issues. They highlight the need to move beyond traditional 'schooling' practices and formulate five principles for designing sustainable development teaching: (1) create engagement for the content of teaching, (2) use the right focus for teaching, (3) deal with local sustainability problems, (4) stress pluralism and (5) include ethical and political dimensions. The authors explain how these principles are based on a pragmatist, transactional didactic theory that understands learning and teaching in terms of action. In this theory, learning is approached as a process of meaning-making that takes into account both students' prior experiences and the specificity of a particular learning situation and that results in a more developed and specific repertoire for coordinating activities with the surrounding world. Learning, then, is about extended possibilities to act, and a vital aspect of it is inquiry, which involves both action and reflection as inseparable activities. This perspective on teaching and learning, the authors argue, makes it possible to stress pluralism and the ethical and political dimension of sustainability issues without falling into an anything-goes attitude.

Part II: choosing teaching content and approaches

The second part of the book is concerned with how to prepare courses and lessons and introduces tools for reflection about teaching practices and the multiplicity of approaches to addressing the ethical and political dimension of sustainability education. The insights, models and typologies offered here are aimed to facilitate the choice of education goals and learning outcomes, subject matter, suitable teaching methods, etc.

In chapter 4, Katrien Van Poeck and Leif Östman present a model for how to plan sustainability teaching by taking into account three functions of education:

qualification, socialisation and person-formation. Qualification is about equipping students with knowledge, skills and understandings that prepare them for a specific task. Socialisation means transferring certain values, attitudes, norms and worldviews in line with the prevailing standards of a particular community or tradition. Person-formation is connected to the formation of the self: the cultivation of personality, the process of personal maturation. The latter can take the form identification, i.e. developing a specific identity in relation to qualification and socialisation, or subjectification, i.e. developing mature and independent ways of being and acting in the world. The authors emphasise and illustrate how qualification, socialisation and person-formation are intertwined in educational practice. Every teaching, they argue, has a certain meaning-making in the forefront, while there are always other meanings that follow automatically, in the background: companion meanings. Hence, teaching is always a value-laden activity. All knowledge content simultaneously and inevitably offers students particular ways of reasoning about the world and their place in it while omitting other possible perspectives and worldviews.

Chapter 5, written by Johan Öhman and Leif Östman, presents different teaching traditions in ESE. Drawing on a large-scale empirical analysis in schools, the authors explain that there exists a variety of ways of teaching about sustainability issues which can be viewed as different 'selective traditions'. Each tradition represents different answers as to what constitutes good sustainability teaching. The authors identified three selective traditions within environmental and sustainability education: a fact-based tradition, a normative tradition and a pluralistic tradition. They describe the differences between the three traditions when it comes to their sustainability approach, didactic approach, approach to facts and values and approach to democracy and education. The purpose of clarifying these traditions here is to establish a reference point that can be applied when discussing teaching that involves issues relating to the environment and sustainable development. The traditions can be seen as alternatives to reflect on, oppose or support when planning lessons or formulating ideas. The strengths and shortcomings of each tradition are discussed in relation to two interconnected premises: that environmental and sustainability issues involve both facts and values and that they should be dealt with democratically.

In chapter 6, Johan Öhman and Leif Östman present and describe a typology of how ethics and morals appear in educational practice: the ethical tendency typology. It distinguishes three different kinds of situations: personal moral reactions, norms for correct behaviour and ethical reflections. A moral reaction is a spontaneous, unpremeditated and non-intentional reaction to a personal experience of what one ought to do or not do in a specific situation. Norms for correct behaviour are social conventions regarding the correct way of acting in certain kinds of social activities. Ethical reflections, finally, are rational and systematic reflections about the reasons for moral actions. The authors illuminate each of the three situations with empirical examples in terms of the diverse learning conditions they bring about. They also discuss the typology as a didactic tool for teachers to use when facilitating students' learning and guiding their moral growth in relation to environmental and sustainability issues. Relating the typology to the pluralistic approach to ESE, they suggest principles for dealing with the ethical tendency in a democratically responsible way and for promoting students' critical thinking and democratic action competence.

Chapter 7 is written by Johan Öhman and David O. Kronlid. First, it presents a pragmatist perspective on ethics and morals based on the works of John Dewey. In this perspective, morals are not innate or fixed, but are something that we learn. This is a continuous process throughout life. We learn by experiencing moral situations that make us reflect on responsibilities and concerns that we have previously taken for granted. In this way, it is argued, we gradually learn to be more sensitive to the specific circumstances that prevail in diverse moral situations and develop an intelligent sympathy. Furthermore, the authors explain how Dewey holds democracy as a moral ideal. In his view democracy is way of life in which people with different experiences create new possibilities by influencing each other. The second part of the chapter suggests two teaching principles based on the pragmatist perspective: (1) start in students' moral experiences of concrete cases and (2) introduce ethical theory and language. These principles provide guidelines for teachers to organise their teaching systematically in a way that supports a moral learning in line with the normative competency necessary for achieving the UN's Sustainable Development Goals.

In Chapter 8, Michael Håkansson, Katrien Van Poeck and Leif Östman present and describe a typology of different ways in which the political dimension of sustainability issues can appear in educational practice. Their political tendency typology distinguishes between political norms, political reflections, political deliberation (sub-divided into normative, consensus-oriented and conflict-oriented deliberation) and the political moment. These categories are explained and illustrated with concrete examples from educational practices. The authors emphasise the different character of each of these situations as well as the very diverse conditions for learning that they bring about. More specifically, they explain how the different categories differ largely as to the opportunities offered for students to develop political action competence when it comes to handling dissonant and conflicting voices in deliberation and decision-making regarding sustainability issues.

In Chapter 9, Ásgeir Tryggvason and Johan Öhman discuss two different approaches to the political dimension of environmental and sustainability education: deliberation and agonism. Discussions about environmental and sustainability issues in classrooms can bring different political visions, opinions and conflicts to the fore. From a pluralistic perspective on ESE, the authors argue, such political differences and conflicts can be seen as a suitable starting point for teaching, rather than as an obstacle to overcome. But how can teachers approach this political dimension of ESE? Deliberation and agonism are outlined as two different approaches to the political dimension of ESE. With a deliberative approach, the role of rational and respectful communication is underscored, as is the ideal to reach for consensus in classroom discussions. With an agonistic approach, the role of emotions and how they are intertwined with political visions in sustainability issues are highlighted. From an agonistic perspective, the teacher should not aim for a consensus in classroom discussions, but instead aim at enabling conflicts and pluralism to have a democratic outlet in discussions. A main point that the authors point to is that deliberation and agonism should be seen as two different approaches to the political dimension in ESE, based on different ideas about classrooms and conflicts and holding different educational consequences.

Part III: *designing and implementing teaching and learning practices*

After having offered the reader insights, models and typologies that can inspire high-quality preparation for teaching sustainable development in Part II, this third part of the book focuses on the *performance* of teaching practices. It brings together a collection of chapters with useful conceptual frameworks, models and typologies about the concrete design and implementation of the prepared practices, which are illuminated by descriptions of practical cases. Part III starts with two chapters (10 and 11) in which the editors elaborate the overall theoretical framework for understanding teaching and learning about sustainable developing that is underlying the subsequent chapters. Whereas Part II provides frameworks, models and typologies that cover a broad range of very diverse ESE practices, Part III deliberately puts forward concepts, frameworks, models and examples that share a pluralistic approach to the ethical and political dimension of teaching sustainable development.

In chapter 10, Leif Östman, Katrien Van Poeck and Johan Öhman introduce a transactional theory of learning. It addresses the specific content that teachers should pay attention to in an ESE context: learning sustainability-related habits that allow for creativity. It also presents models for understanding ESE learning processes: two different 'routes' learning can take (i.e. short and long learning loops) as well as three different 'roots' for learning (i.e. intellectual problems, changes in the physical surroundings and poignant experiences). Furthermore, the authors identify and discuss four crucial factors that influence the learning outcomes: the intrapersonal, the interpersonal, the institutional and the physical. The presented models and factors can be used as a background for designing efficient and fruitful ESE teaching. The chapter ends with a reflection on the relation between bodily feelings and cognition, since an important part of ESE learning concerns not only knowledge but also commitment, standpoints, etc. connected to the ethical and political dimension of sustainability issues.

Chapter 11, also written by Leif Östman, Katrien Van Poeck and Johan Öhman, presents a transactional teaching theory by elaborating how teachers set a scene (learning environment) and stage fruitful inquiries through performing well-considered teacher moves. Teacher moves are interventions made by a teacher to govern the students' learning in a certain direction in line with the purposes of the teaching. There are several types of teacher moves with different functions, for example adding something to the learning environment, confirming or reorienting students' meaning-making, deepening the students' inquiry, etc. Teachers can make epistemological, moral and political moves in order to facilitate students' learning in accordance with specific purposes. Awareness of a wide variety of possible teacher moves, the authors argue, can help teachers to successfully manage the interplay (transactions) between the intrapersonal, interpersonal, institutional and physical aspects that influence students' learning.

In Chapter 12, Katrien Van Poeck, Leif Östman and Johan Öhman go deeper into the concept of ethical moves. They focus on teachers' influence on students' ethical and moral learning and, in particular, on how teachers can promote students' growth as moral subjects in ESE practice. A variety of ethical moves, i.e. actions performed by a teacher that open up a space for articulating moral reactions and for deliberating on ethical opinions, are described and discussed. By performing such moves, the authors argue and illustrate, teachers can turn students' moral experiences into fruitful drivers for pluralistic ESE. It enables students to express and share moral experiences and standpoints, to articulate ethical differences and controversies and to reflect and deliberate on moral reactions and dilemmas.

Chapter 13, written by Katrien Van Poeck and Leif Östman, focuses on teachers' influence on students' learning regarding the political dimension of sustainable development. By employing specific moves, teachers can affect how the political is experienced in ESE. The authors explain and illustrate how political moves can be used in order to give shape to an ESE practice as a conflict-oriented political deliberation in which students raise and defend conflicting standpoints regarding how to organise society in the face of sustainability problems. Different types of political moves are distinguished. Performing them provides opportunities for students to personally experience the judgements, prioritisations and decision-making about different and competing alternatives that are involved in the question of how to organise society sustainably – and to learn from that experience. Teachers can also perform moves that steer the activity away from conflict-oriented deliberation, for example by staging a normative deliberation. Thus, the authors emphasise, different teacher moves affect the way the students experience the political very differently.

Chapter 14 is written by Karin Rudsberg and Johan Öhman. It focuses on students' learning in argumentation that takes place in classroom discussions. The authors first clarify how an argument can be understood as consisting of different elements. Drawing on practical examples, they clarify questions that are important for teachers to think about in relation to students' learning, their use of knowledge and the importance of peer interactions. They conclude that argumentation can be a fruitful method in the teaching of complex, value-related issues. In argumentation, the students not only learn and use content knowledge, but also learn how to formulate valid arguments in order to participate in deliberative discussions. The authors highlight that teachers have an important role to play with regard to the quality and diversity of the deliberation.

In chapter 15, Marie Öhman and Johan Öhman go into the issues of power and governance in ESE classroom practices. They explain how the subject content that is highlighted in school can be understood in terms of power. In Michel

Foucault's view, power is not a question of who has, holds or exercises power. Power is seen as embodied in people's everyday actions, for example the content (knowledge, norms and values) that is offered to students in a teaching situation. The authors explain how the subject content guides students in certain learning directions and thereby favours certain ways of thinking and acting, which in turn create opportunities and restraints for students to understand and look at themselves and their environment in specific ways. The teaching practice in a school subject is often rooted in habits and traditions, and we often regard the content as natural and obvious. By highlighting the power dimension, the authors offer teachers a way of reflecting on the consequences of the choice of content. The chapter is illustrated with examples of ESE classroom practices.

Chapter 16, written by Petra Hansson, proposes four pedagogical principles for staging reading and writing activities in ESE. Together, these principles provide a framework for engaging students in discussions of the sustainability challenges we face today through structured literate activities. This chapter offers an account of the theoretical inspirations to the principles including ecocriticism, reader-response theory and theories of writing and responding. Furthermore, it provides teachers with a practical description of the pedagogical principles using the novel *Robinson Crusoe* as an example of how the teaching of literature can be organised to address sustainability issues as well as to encourage students to express opinions, emotions and values. The author emphasises that activities that focus on expressing and sharing opinions and acknowledging emotions require an equally structured approach as pedagogies that focus on the learning of specific sustainability knowledge and skills.

In chapter 17, Louise Sund and Karen Pashby focus on global issues in the classroom. They argue that schools have an important role to play in preparing students to engage responsibly with global equity and justice in ways that directly take up rather than step over ethical issues. Despite a general policy consensus on the importance of including such issues in education, there is a lack of knowledge about how to deal with them in educational theory and practice. This chapter offers a didactical reflective tool to support teachers in making choices about content and pedagogy grounded in ethical questions and complex understanding of global justice and equity issues. The authors illustrate with some empirical examples that teachers and students can experience a sense of significance and worthiness of engaging in a more critical approach. They argue that if we as teachers critically engage with notions of complexity and complicity while also being rooted in the daily life of classrooms, we open up possibilities for approaches to global issues pedagogy that takes up rather than glosses over the complex set of factors involved in global issues.

Chapter 18, written by Iann Lundegård, offers a glimpse from a classroom in which the teacher aims to organise an authentic education in the sense that it allows the students to act as political subjects. From a didactical perspective, the author describes how a teacher can perform a teaching practice that acknowledges the inevitable link between knowledge and values. Furthermore, with the help of a pragmatist perspective, he argues how these two aspects of the students' meaning-making closely interact. Simultaneously, we get insight in how ESE based on freedom and democracy renders opportunities for the students to declare personal relations to the content at hand, but also how this freedom is ultimately achieved through mutual encounters in-between them.

Chapter 19, written by Pernilla Andersson, points to the need for education to prepare students to make decisions (also) in the absence of clear guidelines and regulation. The chapter focuses on this need by presenting teaching approaches that offer students embodied experiences of decision-making in the face of sustainability problems characterised by uncertainty and complexity. The author provides short practical examples from business education at upper secondary level to illuminate (a) when different worldviews regarding how sustainability problems should or could be addressed come to the fore and (b) emancipatory educational qualities in terms of subjectification. The examples illuminate what could be described as 'dislocatory moments'. Drawing on the concept of 'dislocatory moments', the author presents a didactic model that can be used to identify room for subjectification processes together with a change of views regarding sustainability issues. The ambition is to facilitate teaching that can contribute to change for sustainability without compromising emancipatory education ideals.

In chapter 20, finally, Ásgeir Tryggvason and Andreas Mårdh address the topic of political emotions in educational practice. When discussing sustainability issues in the classroom, heated emotions and conflicts between students can arise. But discussions about sustainability issues can also lack engagement and emotional involvement from the students, even when the teacher brings up what s/he thinks is a burning sustainability issue. A crucial question is then how to approach emotions in ESE in a non-instrumental way, where the political dimension of emotions can be put to the fore. This chapter outlines two strategies to teach with and through political emotions in ESE. The first strategy is simplification of the complexity and the conflictual aspect of sustainability issues. When a classroom discussion is characterised by an emotional indifference or a lack of engagement, simplification is a strategy to approach this indifference. The second strategy is circulation, which is a way to maintain the intensity of emotions, or to (re)orientate them toward other objects and issues. When emotions run high in a discussion, circulation is a strategy to approach these emotions as productive elements of a vibrant environmental and sustainability education.

References

Ashley, M. (2000) "Science: An unreliable friend to environmental education?" *Environmental Education Research*, 6(3): 269–280.

Håkansson, M., Kronlid, D.O. and Östman, L. (2017) "Searching for the political dimension in education for sustainable development: Socially critical, social learning and radical democratic approaches". *Environmental Education Research*, pre-published online: https://doi.org/10.1080/13504622.2017.1408056

Knutsson, B. (2013) "Swedish environmental and sustainability education research in the era of post-politics?" *Utbildning & Demokrati*, 22(2): 105–122.

Lundegård, I. and Wickman, P.-O. (2007) "Conflicts of interest: An indispensable element of education for sustainable development". *Environmental Education Research*, 13(1): 1–15.

Öhman, J. ed. (2008) *Values and Democracy in Education for Sustainable Development: Contributions From Swedish Research*. Liber, Malmö.

Östman, L. (2010) "Education for sustainable development and normativity: A transactional analysis of moral meaning-making and companion meanings in classroom communication". *Environmental Education Research*, 16(1): 75–93.

Sund, L. and Öhman, J. (2014) "On the need to repoliticise environmental and sustainability education: Rethinking the postpolitical consensus". *Environmental Education Research*, 20(5): 639–659.

United Nations Educational, Scientific and Cultural Organization – UNESCO. (2014a) *Sustainable Development Begins With Education: How Education Can Contribute to the Proposed Post-2015 Goals*. UNESCO, Paris.

United Nations Educational, Scientific and Cultural Organization – UNESCO. (2014b) *Roadmap for Implementing the Global Action Programme on Education for Sustainable Development*. UNESCO, Paris.

Van Poeck, K. (2018) "Environmental and sustainability education in a post-truth era. An exploration of epistemology and didactics beyond the objectivism-relativism dualism". *Environmental Education Research*, pre-published online: https://doi.org/10.1080/13504 622.2018.1496404

Van Poeck, K., Östman, L. and Block, T. (2018) "Opening up the black box of learning-by-doing in sustainability transitions". *Environmental Innovation and Societal Transitions*, pre-published online: https://doi.org/10.1016/j.eist.2018.12.006

Part I

Education and the challenge of building a more sustainable world

1 Four misunderstandings about sustainability and transitions

Thomas Block and Erik Paredis

Introduction

'Sustainability' and 'transition' are popular and often-used concepts. However, their omnipresence also leads to a diversity of interpretations and misunderstandings about their meaning. This can be annoying in a teaching situation: what do you tell students when they ask for a definition? And how do you make these concepts concrete in educational projects? In this chapter, we argue that the diversity of interpretations is a consequence of the political character of these concepts: different societal groups attach so much importance to sustainability and transitions that they want to influence the debate in their favour and consequently propose definitions that fit with their ideas or serve their interests. Interestingly, in that way concepts such as sustainability and transitions open up a space for a societal debate about the direction our societies should take and the choices that have to be made. We examine four common misunderstandings about sustainable development and sustainability transitions: (1) 'Sustainability is about ecological concerns'; (2) 'We need a waterproof and objective definition of sustainability'; (3) 'Every change leads to a transition'; and (4) 'We can easily plan and manage sustainability transitions'. Obviously, the arguments we develop should be considered as part of an on-going discussion about sustainability transitions. But we hope that by drawing some lines, we can contribute to the usefulness of these concepts for education and teaching.

Misunderstanding 1: 'sustainability is about ecological concerns'

Of course, sustainability is about ecological issues, but that's only one part of the story. In particular in industrialised societies, the discussion about sustainability tends to be narrowed down to the ecological domain, while the challenge lies essentially in how we can develop societies that provide a high quality of life in a way that is simultaneously ecological and just, not only locally or nationally, but also globally. In fact, this was already part of the most influential definition of sustainable development, namely the one dating from the report *Our Common Future*, published in 1987 by the United Nations (UN). This so-called Brundtland Report defines sustainable development as 'a development that meets the needs of the present without compromising the ability of future generations to meet their own needs' (WCED 1987, 8). Meeting needs obviously does not only refer to a clean environment, but also to food, housing, education and so on. This interpretation of sustainability as implying both the social and ecological domain has been reaffirmed in 2017 in the 17 Sustainable Development Goals (SDGs) that were approved by the UN (2017). The SDGs cover a broad spectrum of environmental, social, economic and institutional goals.

So what does it mean that the challenge of sustainability lies essentially in the combination of social and ecological goals? We start by briefly discussing them separately. One of the most influential interpretations of the ecological challenge derives from Rockström et al. (2009), who introduced the planetary boundary concept, which defines the environmental limits within which humanity can safely operate. An updated analysis of this framework (Steffen et al. 2015) shows that four of nine planetary boundaries have now been crossed as a result of human activity: climate change, loss of biosphere integrity, land-system change and altered biogeochemical cycles (see Figure 1.1) These four exceedances also have a pervasive influence on the remaining boundaries.

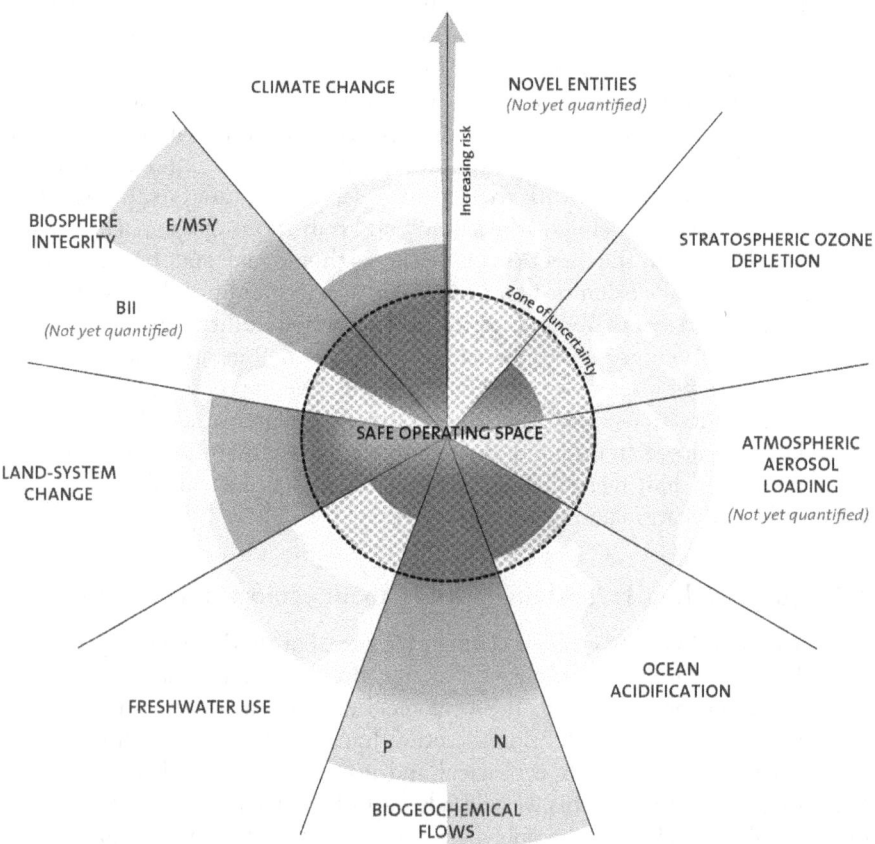

Figure 1.1 Estimates of how the different control variables for seven planetary boundaries have changed from 1950 to present

Source: (Steffen et al. 2015)

What about the social side of sustainability? While the number of people living in extreme poverty dropped by more than half between 1999 and 2013 – from 1.7 billion to 767 million (UN 2017) – too many are still struggling for the most basic human needs such as food, water, shelter, education and health care. Moreover, authors such as Wilkinson and Pickett (2009), Piketty (2014) and Alvaredo et al. (2018) have documented the growth of inequality and have shown that income inequality has increased in nearly all world regions in recent decades. According to the Credit Suisse Research Institute (2018) and as shown in Figure 1.2, the richest decile (top 10% of adults) owns 85% of global wealth, and the globe's richest 1% owns almost half the world's wealth, while the bottom half collectively owns less than 1% of total wealth.

Such social and ecological concerns are challenging in themselves, but a typical characteristic of sustainable development lies in the combination of the two. Sustainability is about ecological and social concerns at the same time. An interesting way to assess sustainability is, for instance, by combining the ecological footprint (EF) for the ecological concerns with the human development index (HDI) for the social concerns. The EF relates to the impact of human activities measured in terms of the area of biologically productive land and water required to produce the goods consumed and to assimilate the wastes generated. An EF of less than 1.7 global hectares (gha) per person makes the resource demands globally replicable (this is sometimes called 'the fair earth share'), but the average per capita is currently 2.7 gha per person. So the average EF per person worldwide needs to fall significantly, especially in Northern countries where the EF is often higher than 7 gha per person. The human development index is an index that measures key dimensions of human development: life expectancy, years of schooling and gross domestic product per capita. The UN considers an HDI over 0.8 to be 'very high human development'. With these two thresholds, we can define two minimum criteria for global sustainable development, namely an average EF lower than 1.7 gha per person and an HDI of at least 0.8. Figure 1.3 shows that only a few countries come close to achieving sustainability in this definition.

This figure also makes clear what the main sustainability challenges are: (1) for countries in the left lower corner: creating a high level of human development

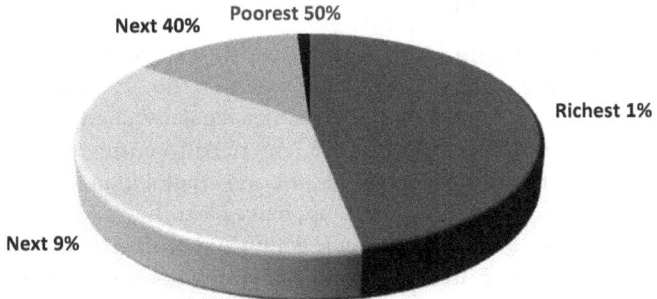

Figure 1.2 Division of global wealth

Source: (based on Credit Suisse Research Institute 2018)

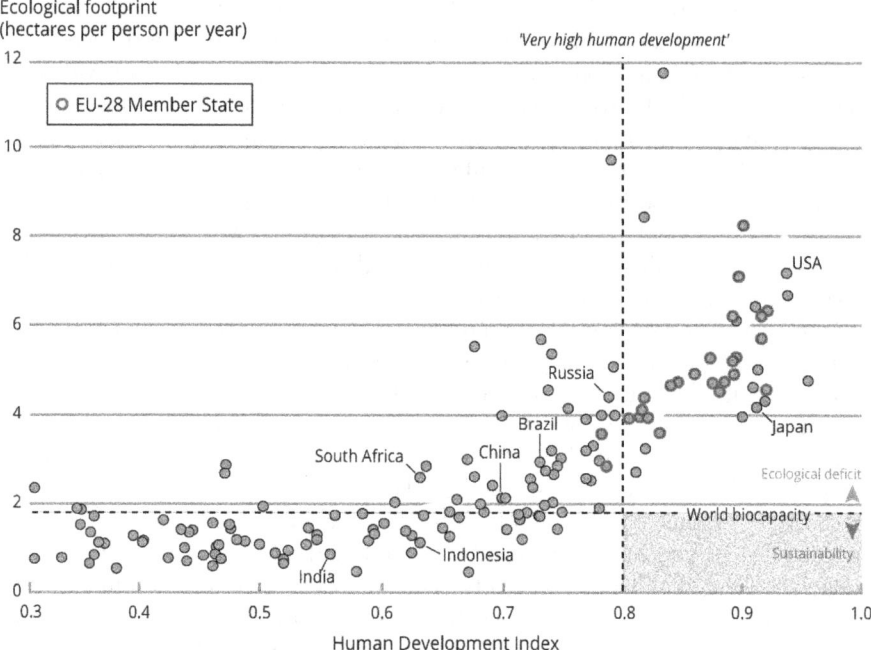

Figure 1.3 Correlation of ecological footprint and the human development index
Source: (European Environment Agency 2015; Global Footprint Network)

without depleting the planet's or a region's ecological resource base, and (2) for countries in the upper right corner: creating a lower ecological footprint while retaining high human development. This global idea of sustainability as a combination of quality of life, ecological limits and justice can also be used to define the rough contours of sustainability within a country, a city, a neighbourhood, a project, a course, etc.

Misunderstanding 2: 'we need a waterproof and objective definition of sustainability'

A fixed definition of sustainability ignores the political character and the custom-made approach this guiding concept requires. Even the oft-cited definition from the Brundtland Report (see above), or another regularly cited definition such as the one of Agyeman et al. (2003, 5), i.e. 'the need to ensure a better quality of life for all, now and into the future, in a just and equitable manner, whilst living within the limits of supporting ecosystems', make clear that 'sustainability' is based upon normative principles such as needs, equity and ecological limits that cannot be unequivocally defined. Instead, the sustainability concept is shaped, used and 'owned' by an ever-widening range of stakeholders (Hopwood et al. 2005). This democratisation of sustainability is a positive evolution as it testifies to the power of attraction and the enduring relevance of the concept (Hugé et al. 2016). But a consequence is that the diversity of stakeholders engaging with sustainability (e.g. academics, governments, business, NGOs) gives rise to a multitude of interpretations, because actors try to produce interpretations that favour their interests.

Inspired by an article of Hopwood et al. (2005) of 14 years ago, we mapped different interpretations of sustainable development (see Figure 1.4). Hopwood et al. use a socio-economic and an environmental axis to distinguish between three views on the nature and scope of change, ranging from status quo to reformist to transformation.

Views in the status quo band aspire to a belief in 'business as usual': sustainability can be achieved within existing political structures and economic growth models. Growth provides resources to pay for environmental measures and technologies, and 'trickle-down economics' will solve the poverty problems (as in this view economic benefits for the wealthy will trickle down to the poorer members of society). This perspective counts on eco-efficiency, international competitiveness and free markets (Hopwood et al. 2005). The challenge essentially lies in finding the right combination of technologies to meet rising demands in sustainable ways. Within status quo approaches the market is often the agent of transformation, through which pricing, creating markets and property rights regimes unleash new rounds of 'green accumulation' (Scoones et al. 2015). Reformist models are critical of current policies of governments and business, but they still believe that sustainability problems can be solved within the current political and economic structures. Green economists support capitalism, but give the government an important role (Hopwood et al. 2005). The starting point is often the need to re-embed (global) markets in stronger frameworks of social

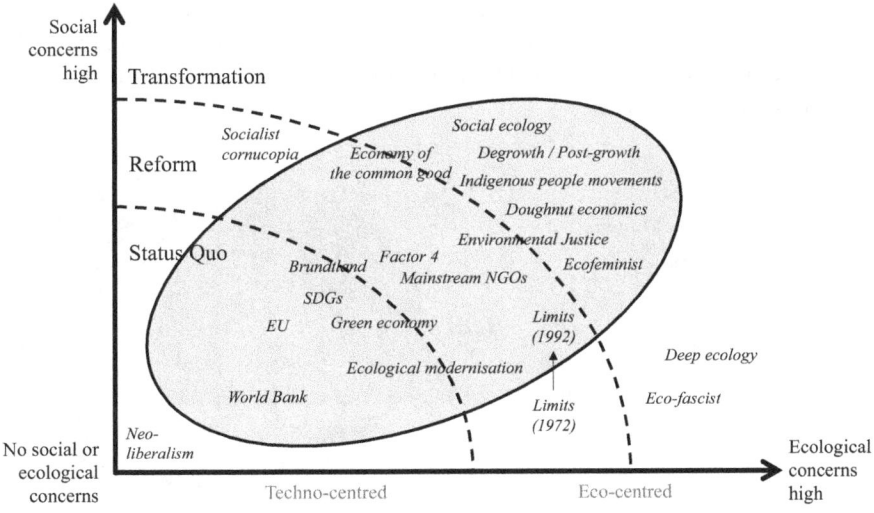

Figure 1.4 Mapping of views on sustainable development

Source: (based on Hopwood et al. 2005)

control, combined with a recognition of the state's historically central role in previous waves of innovation and financing of technology and growth (Scoones et al. 2015). According to transformative approaches, a fundamental change is needed because sustainability problems 'are located within the very economic and power structures of society' and in 'how humans interrelate with the environment' (Hopwood et al. 2005, 45). Therefore, approaches such as degrowth or doughnut economics postulate that besides eco-efficiency, also sufficiency, redistribution and decommodification are essential sustainability strategies (D'Alisa et al. 2014). Transformative approaches suggest that transformations often arise from political action of groups outside the centres of power (indigenous groups, women, the poor and working class).

Over the years, the mainstream of the debate – mainly carried by UN organisations, national governments, the EU, World Bank, international NGOs, etc. – has been a mixture of the status quo view and the reform view (Hopwood et al. 2005; Paredis 2013). This fits perfectly within the idea of ecological modernisation, an approach that argues that the capitalist economy benefits from moves towards environmentalism. Through a smart interaction of technology, science, business and markets, with a government that intervenes by setting standards and providing incentives through market mechanisms, ecological modernisation promises a 'positive-sum game', where economic growth and ecological protection can be combined (Hajer 1995; Paredis 2013).

As stated in the introduction, we argue that the diversity of interpretations is a consequence of the political character of the concept of sustainability. But does that mean that all interpretations are of equal merit? Although we do not advocate a fixed definition of sustainability, neither are we in favour of 'anything-goes' relativism. For us, the concept of sustainability opens up a space to think beyond status quo views, and as such beyond the ideas of 'technological fix' and 'trickle-down economics'. In this respect, we propose to use the nested model to visualise the essence of sustainability and not the iconic figure of the triple bottom line (see Figure 1.5).

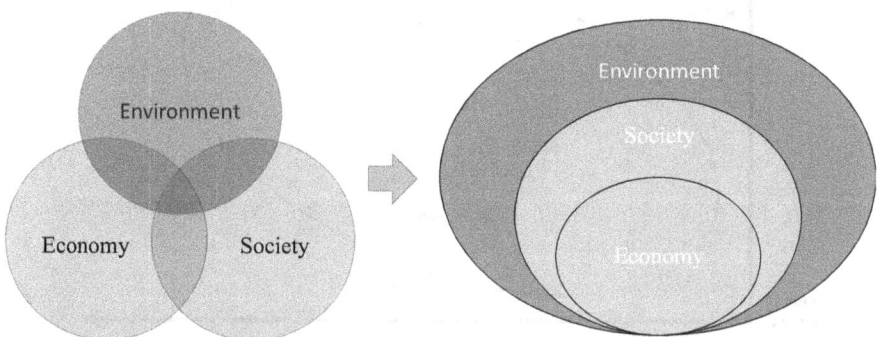

Figure 1.5 From the iconic figure of the triple bottom line (left) towards the nested hierarchy model (right)

Source: Author's own.

The triple bottom line, also expressed in the phrase 'people, planet and profit', considers 'business' and 'profit' as goals in itself, next to social and ecological challenges, and focuses on the mutual maximisation of the three dimensions. The win-win assumption behind this model neglects the obvious link between current economic activities and planetary environmental degradation (Isil and Hernke 2017) or the growth of inequality. The nested hierarchy model applies a logic from ecological economics that takes as its basis a socially discussed ecological scale which subsequently determines the contours of a fair distribution and only then leaves room for the market to determine how to realise an efficient allocation (Daly 1992; Martínez-Alier and Muradian 2015).

This plea for taking reform and transformative approaches seriously implies automatically a call for fundamental changes in societal structures, practices and ways of thinking. For understanding such changes, we appeal to transition studies, a discipline which is also confronted with some misunderstandings.

Misunderstanding 3: 'every change leads to a transition'

The word transition has become omnipresent in discussions about more sustainable societies, but not all change can be equalled to a transition. A transition is not just a question of a new technological product, a new policy rule or a specific behavioural change. A transition is usually defined as a deep, fundamental change in a societal system, such as the energy system, the mobility system or the food system. Transitions are long-term processes in which existing structures, cultures and practices that are anchored in a society are broken down and new ones become dominant. During a transition, a system changes in multiple dimensions: technology, actors, rules, infrastructures, power relations, patterns of thinking, problem definitions and solutions, cultural meanings, etc. (Geels 2007; Loorbach 2007).

More precisely, the literature on sustainability transitions holds that a transition is a radical change in the regime of a system. What is a regime? A regime is the dominant way of fulfilling a societal function, consisting of the dominant technologies, actor networks, rules, practices, artefacts, infrastructures, ways of thinking. In the mobility system, for example, the regime is centred around the car, with elements such as (Geels 2012) (a) an industry investing in the internal combustion engine, (b) dense networks of roads, highways and other spatial infrastructures (e.g. petrol stations and parking), (c) user patterns and cultural values related to 'my car, my freedom', (d) vested interests (industry, car lobby) that resist major change and (e) government rules and regulations, e.g. on companies cars, fuel taxes, traffic regulations and urban planning (allowing urban sprawl). The interconnectedness of all these elements keeps the current automobility regime dominant and stable. They function as lock-in mechanisms, 'leading incumbent actors to prefer incremental changes that stay within the bounds of the existing regime' (Geels 2012, 478). A change in one of these elements – such as the introduction of an electric car – is not enough to speak of a transition. Transitions happen along multiple dimensions and do not result from one single driver or cause, but instead involve co-evolutionary developments between all those dimensions (Geels 2012). A regime change is hard and complex, and consequently difficult to steer towards a sustainable future.

Misunderstanding 4: 'we can easily plan and manage sustainability transitions'

From the previous paragraphs, it will be obvious that transitions cannot simply be managed. A sustainability transition is the result of complex interactions of different processes and factors. A useful framework to analyse and to understand transitions is the multi-level perspective (MLP). The basic premise of the MLP is that transitions are non-linear processes that result from the interplay of multiple developments at three analytical levels: (1) niches (the locus for radical innovations), (2) socio-technical regimes (the locus of established practices and associated rules) and (3) an exogenous socio-technical landscape (Geels 2007, 2012; Geels and Schot 2007).

We already discussed regimes as the dominant way of fulfilling a specific societal function. While change at the regime level is incremental, at the micro-level radical novelties emerge. In these niches, technologies and practices are developed that diverge strongly from what is normal in the regime. Niches can be constructed around technologies, new practices or approaches to governance and policy (Paredis 2013). Because a lot of experimentation is going on and technologies, rules and practices are in the making, niche configurations are less stable than regimes. Using our example of automobility, promising niche developments are battery-electric vehicles or fuel-cell vehicles; car sharing or public bikes; workplace and school travel plans; special bus lanes; sustainable urban planning with car-free zones. The macro-level is called the socio-technical landscape and refers to 'the technical, physical and material backdrop that sustains society' (Geels and Schot 2007, 403). The landscape usually evolves rather slowly and contains deep cultural patterns, macro-political developments, natural circumstances and material infrastructure. In mobility, we can consider climate change, peak oil and technology diffusion as landscape trends that are important sources of pressure on the existing automobility regime. But there are also societal characteristics and developments that help stabilise the regime: preferences for private property, feelings of autonomy or the separation of work and home (Geels 2012).

The central insight of the MLP is that transitions can be understood as a complex interaction between niche, regime and landscape factors (see Figure 1.6). A regime can become disrupted under the pressure of landscape developments, internal regime problems (such as traffic jams and air pollution in our example) and niche-innovations that provide alternatives. Particularly when different types of pressure occur in conjunction, windows of opportunity can arise that enable niches to break through and cause radical change in the regime, with new practices and technologies and new actors that become dominant. However, there is no guarantee that transitions succeed: niche-innovations may fail to build up sufficient momentum or suffer setbacks, or tensions in existing regimes

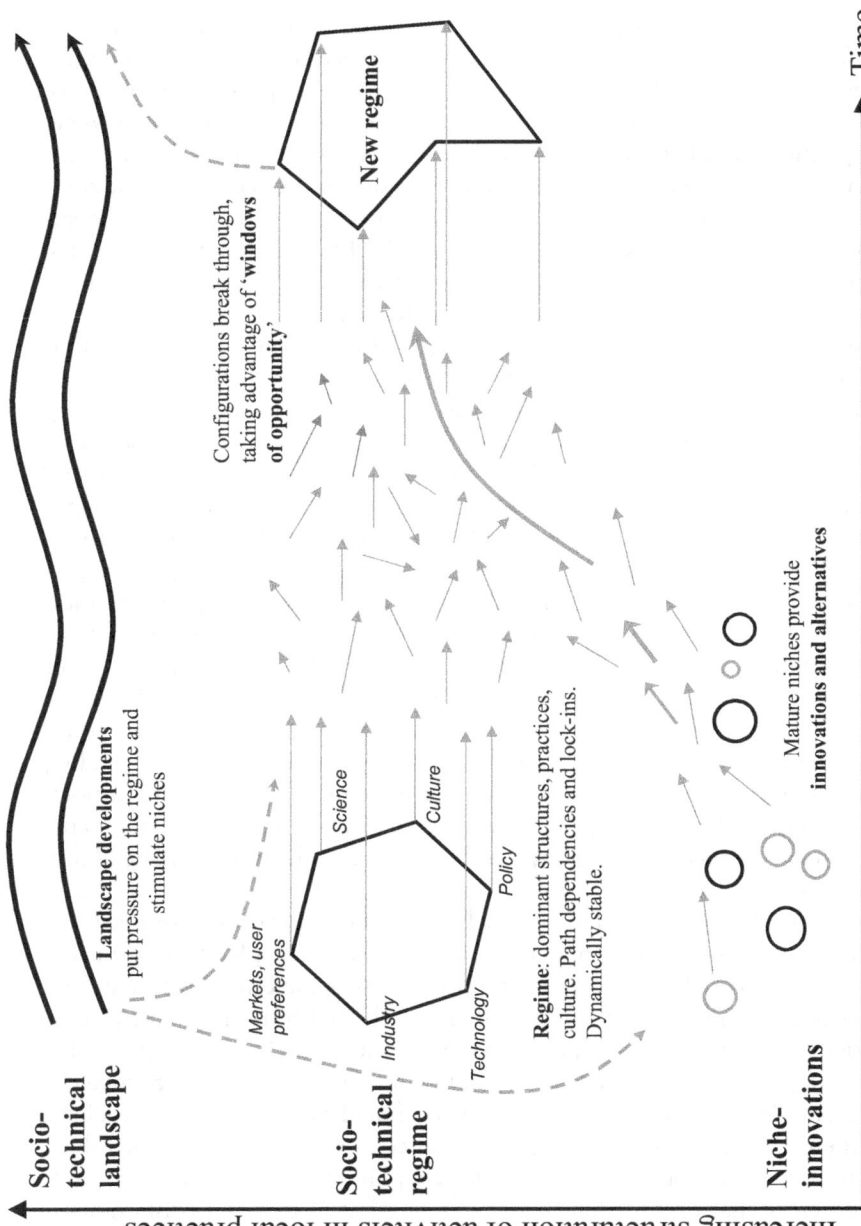

Figure 1.6 The multi-level perspective

Source: (based on Geels and Schot 2007, 401)

may remain small so that 'windows of opportunity' for niche-innovations do not materialise (Geels 2007, 2012).

Then, is it possible to plan and manage transitions? Most transition scholars agree that the answer is 'no', and certainly not in the early phases of a transition. The processes are too complex for simply managing a transition. But influencing may be possible, in particular when actors succeed in intensifying pressures on current unsustainable regimes. Different strategies are possible, for example developing experiments and niches that can serve as alternatives to the regime. Or exposing contradictions and problems in the regime. Or trying to influence consumer practices and the daily routines of consumers. Remarkably, some transition approaches occasionally seem somewhat naïve in how they treat the complexity and politics involved in transitions, and in that way they risk reinforcing the idea of straightforward management. In particular, the approach of 'transition management' (TM) has received criticisms in that vein, and not only because the connotations of the word 'management' suggest the presence of managers that have clear-cut solutions and the ability to steer a system in a linear way.

Transition management, developed by Rotmans, Loorbach and Kemp (Kemp and Loorbach 2006; Rotmans and Loorbach 2010), aims to facilitate and accelerate sustainability transitions. TM usually employs a so-called transition arena with a selected number of frontrunners in the system, from the government, business, civil society and science. Such arenas are intended to be 'relatively safe and free, protected environments without any power hierarchy' (Rotmans and Loorbach 2010, 218). This mixed network of actors develops a transition agenda, which contains a commonly developed system analysis, a future sustainability vision for the system, transition paths towards that vision and a series of experiments to test and initiate the paths in reality. The underlying rationale is one of 'goal-oriented incrementalism' (Rotmans and Loorbach 2010): transition management is done with a long-term desirable goal or perspective as guiding image, but working towards it will have to be done in small steps. Furthermore, goals are not fixed but developed through processes of learning and experimenting, and the process of working towards them requires adaptation and flexibility (Paredis 2013).

The main problem of the TM approach is that it has no political strategy outside the 'power free' transition arena. It acknowledges complexity as a starting point, but when applied in practice the complexity is too often neglected. The TM approach is definitely useful to create networks of frontrunners, to create an ambitious sustainability vision, to couple interesting practices and experiments and, as such, in trying to influence the political agenda, but how to change the regime and deal with lock-in mechanisms is underdeveloped. This is problematic when a TM process wants to gain influence in policy. The actors involved have to show active agency that looks for couplings with contemporary trends and processes, that tries to change regime rules, that searches for confrontation with dominant discourses and that engages with institutionalisation (Paredis 2013). Neither the theory nor the practical guidelines of TM have anything to say about this kind of agency. The risk is that a 'power-free arena' goes hand in hand with a 'powerless network', and it shows the necessity to move away from TM as a stand-alone approach and embed it as part of a broader transition governance strategy.

The politics of sustainability and transitions

The lack of clear definitions of words such as 'sustainability' and 'transitions' can at first sight seem problematic, but when interpreted as part of a political struggle about the future of our societies, it opens up a space for debate. In this 'reconstructive exercise' (Sneddon et al. 2006, 264), sustainable development is thought of as a framework that allows debating the fundamental choices humanity is facing: what type of society do we want to sustain, or how do we live the good life within the safe and just operating space for humanity (Boström and Davidson 2018; Raworth 2017)? This approach acknowledges 'the inherently normative and political nature of sustainability' (Robinson 2004, 381), even takes it as point of departure or 'embraces' it (Sneddon et al. 2006), and thus devotes as much attention to the process characteristics of sustainable development as to its content (Paredis 2013). Grin et al. (2010, 2) see sustainable development 'as an open-ended orientation for change'. Such open-endedness is considered as a strength since it allows pluralistic appropriation in continuous, deeply political and participatory processes driven by different stakeholders. As we know from transition research, these deeply political processes are characteristic of the changes that occur during transitions. Transforming regimes in, for example, the energy, mobility or food system always meet with power struggles, conflicts of interests and resistance. Incumbent actors try to safeguard their position while niche actors try to gain a foothold. Invariably, controversies emerge over how the transition should be brought about, who the responsible actors are, who will win and who will lose, or what the relevant technological and scientific pathways are. As mentioned above (see Misunderstanding 2: 'We need a waterproof and objective definition of sustainability'), proponents of technology-led, marketised, state-led or citizen-led transformations suggest different frames, different politics, different strategies, different alliances between actors, and so different routes to achieving transformations, often advocating a combination of pathways and demonstrating that there is no one-size-fits-all approach to sustainability transitions and their politics (Scoones et al. 2015). As such, it is necessary to ask what is put on the agenda, what is decided, by whom, to what ends and with what effects.

Summarising, we can say that the misunderstandings that regularly surface in the debate about sustainability and transitions are in fact unavoidable. They are inherently linked to the fact that many groups in society attach importance to these concepts and try to influence the discussion. Instead of interpreting this as a problem, we have to learn to deal with it and turn it into a chance for a broad societal debate. And, relevant for this book: this also offers chances for teaching. Finally, the fact that sustainability processes are deeply political also leads to specific challenges for knowledge production and education at universities (see chapter 2).

References

Agyeman, J., Bullard, R. and Evans, B. eds. (2003) *Just Sustainabilities: Development in an Unequal World*. MIT Press, Cambridge, MA.

Alvaredo, F., Chancel, L., Piketty, T. et al. (2018) *World Inequality Report 2018*. World Inequality Lab.

Boström, M. and Davidson, D.J. eds. (2018) *Environment and Society: Concepts and Challenges*. Palgrave Macmillan, London.

Credit Suisse Research Institute. (2018) *Global Wealth Report 2018*. Credit Suisse, Zurich.

D'Alisa, G., Demaria, F. and Kallis, G. (2014) *Degrowth: A Vocabulary for a New Era*. Routledge, London.

Daly, H.E. (1992) "Allocation, distribution and scale: Towards an economics that is efficient, just and sustainable". *Ecological Economics*, 6: 185–193.

European Environment Agency. (2015) (www.eea.europa.eu/data-and-maps/figures/correlation-of-ecological-footprint-2008). Accessed 24 April 2019.

Geels, F.W. (2007) "Transformations of large technical systems: A multi-level analysis of the Dutch highway system (1950–2000)". *Science Technology & Human Values*, 32(2): 123–149.

Geels, F.W. (2012) "A socio-technical analysis of low-carbon transitions: Introducing the multi-level perspective into transport studies". *Journal of Transport Geography*, 24: 471–482.

Geels, F.W. and Schot, J.W. (2007) "Typology of sociotechnical transition pathways". *Research Policy*, 36(3): 399–417.

Grin, J., Rotmans, J. and Schot, J. eds. (2010) *Transitions to Sustainable Development: New Directions in the Study of Long Term Transformative Change*. Routledge, New York.

Hajer, M. (1995) *The Politics of Environmental Discourse: Ecological Modernization and the Policy Process*. Clarendon Press, Oxford.

Hopwood, B., Mellor, M. and O'Brien, G. (2005) "Sustainable development: Mapping different approaches". *Sustainable Development*, 13: 38–52.

Hugé, J., Block, T., Waas, T., Wright, T. and Dahdouh-Guebas, F. (2016) "How to walk the talk? Developing actions for sustainability in academic research". *Journal of Cleaner Production*, 137: 83–92.

Isil, O. and Hernke, M.T. (2017) "The triple bottom line: A critical review from a transdisciplinary perspective". *Business Strategy and the Environment*, 26: 1235–1251.

Kemp, R. and Loorbach, D. (2006) "Transition management: A reflexive governance approach". In Voß, J.-P., Bauknecht, D. and Kemp, R. eds., *Reflexive Governance for Sustainable Development*. Edward Elgar, Cheltenham, 103–130.

Loorbach, D. (2007) *Transition Management, New Mode of Governance for Sustainable Development*. International Books, Utrecht.

Martínez-Alier, J. and Muradian, R. (2015) "Looking forward: Current concerns and the future of ecological economics". In Martínez-Alier, J. and Muradian, R. eds., *Handbook of Ecological Economics*. Edward Elgar, Cheltenham, 473–482.

Paredis, E. (2013) *A Winding Road: Transition Management, Policy Change and the Search for Sustainable Development*. UGent, Gent.

Piketty, T. (2014) *Capital in the Twenty-First Century*. Harvard University Press, Cambridge, MA.

Raworth, K. (2017) *Doughnut Economics: Seven Ways to Think like a 21st-Century Economist*. Random House, London.

Robinson, J. (2004) "Squaring the circle? Some thoughts on the idea of sustainable development". *Ecological Economics*, 48: 369–384.

Rockström, J., Steffen, W., Noone, K., Persson, Å., Chapin, F.S. et al. (2009) "A safe operating space for humanity". *Nature*, 461: 472–475.

Rotmans, J. and Loorbach, D. (2010) "Towards a better understanding of transitions and their governance: A systemic and reflexive approach". In Grin, J., Rotmans, J. and Schot, J. eds., *Transition to Sustainable Development: New Directions in the Study of Long Term Transformative Change*. Routledge, New York, 105–220.

Scoones, I., Leach, M. and Newell, P. (2015) "The politics of green transformations". In Scoones, I., Leach, M. and Newell, P. eds., *The Politics of Green Transformation*. Routledge, Abingdon, 1–24.

Sneddon, C., Howarth, R.B. and Norgaard, R.B. (2006) "Sustainable development in a post-Brundtland world". *Ecological Economics*, 57: 253–268.

Steffen, W., Richardson, K., Rockström, J. et al. (2015) "Planetary boundaries: Guiding human development on a changing planet". *Science*, 347(6223): 1259855.

United Nations. (2017) *Sustainable Development Goals*. (https://sustainabledevelopment. un.org/sdgs). Accessed 24 April 2019.

WCED (1987) *Our Common Future*. Oxford University Press, Oxford.

Wilkinson, R. and Pickett, K. (2009) *The Spirit Level: Why More Equal Societies Almost Always Do Better*. Allen Lane, London.

2 Tackling wicked problems in teaching and learning. Sustainability issues as knowledge, ethical and political challenges

Thomas Block, Katrien Van Poeck and Leif Östman

Introduction

In this chapter we elaborate how, due to some specific features that characterise many sustainability problems, sustainable development teaching does not only involve questions of knowledge but also ethical and political challenges. Starting from concrete examples and a typology of sustainability problems, we describe how factual uncertainty (scientific controversy) and disagreement on values and norms (ethical and political controversy) bring about specific challenges for knowledge production about sustainability problems and possible solutions. Scholars have therefore proposed new approaches to science such as 'post-normal science'. Seeking inspiration in these proposals, we articulate the ethical and political challenges involved in sustainable development teaching and explore what it could mean to design 'post-normal education' for dealing with wicked sustainability problems where facts are uncertain, values in dispute, stakes high and decisions urgent.

The wickedness of sustainability problems

Find the differences: the ozone hole, climate change and genetically modified food

In the pursuit of sustainability, we face a variety of problems. Three illustrations of sustainability problems exemplify how such issues can vary substantially with regard to (un)certainty about expert knowledge and (dis)agreement on norms, values and solutions.

In the 1980s scientists discovered the annual depletion of ozone above the Antarctic (Farman et al. 1985; Stolarski et al. 1986), followed by an almost immediate scientific consensus about the main cause of the ozone hole, i.e. the use of manufactured chemicals, especially chlorofluorocarbons (CFCs). Politicians didn't hesitate and already in 1987 the Montreal Protocol, an international agreement aiming to phase out CFCs, was drawn up. CFCs were widely used, be it in a limited number of applications: the aerosol industry and refrigeration and air-conditioning systems. After some initial resistance, chemical manufacturers were willing and able to produce substitutes for CFCs. So, we definitely faced a severe and urgent sustainability problem, but the scientific evidence of causes and solutions was considered strong and clear and, moreover, both policymakers and industries were able to envision and implement solutions that did not demand a radical change of our way of life (Grundmann 2016). The production of ozone-depleting gases decreased fast and, although the ozone hole will not immediately disappear as CFCs are stable gases that remain in the atmosphere for decades after release, this once very worrying sustainability problem is today considered as being solved.

The central role of scientific, political and societal consensus is often transferred as a key principle from the ozone case to other sustainability problems such as climate change. Yet, what we face here is a very different kind of problem. The global climate system is much more complex than the ozone hole and, despite strong scientific evidence that global warming is caused by the emission of greenhouse gases, expert knowledge about the consequences of climate change and proper solutions in view of mitigation and/or adaptation is often incomplete, uncertain and contested. Hence, the climate change debate seems impossible to win purely by scientific arguments because it is not, ultimately, a scientific problem with clear and simple solutions (Hulme 2009; Goeminne 2011). It is also a matter of value judgements about which facts matter (most). As the emission of greenhouse gases – especially CO_2 – is inherently entangled with a multitude of practices that constitute people's lifestyles worldwide, we are confronted with a sophisticated, multi-dimensional and multi-layered problem which has as much to do with our human society than with the interplay of sun and atmosphere. We are all caught up in the problem, be it through various kinds of attachments, ranging from the economic interests of the automobile industry over Western consumers' association of materialism with the idea of the good life, to Maldivians' fear of losing their habitat to rising sea levels. Many of the proposed solution pathways are unachievable without a radical change of our way of life. Despite the dominant scientific focus on CO_2 efficiency (Goeminne 2010), the climate change debate shows that our understanding of the problem differs widely. We see it through lenses of our cultural and personal expectations, and we all have different perspectives on the risks (Hulme 2009): some fear environmental or social aspects, other focus on economic or political problems. Hence, factual claims in terms of how much CO_2 we emit often give rise to contestation when weighed against the question of the good life. As such, it seems a problem that cannot be solved, but must instead be resolved and renegotiated, over and over again (Grundmann 2016). In these deliberations, facts and values intertwine in disputes over which facts matter (most).

This is also observable in controversies over genetically modified organisms (GMOs) and other uses of genetic engineering in food production. In Belgium, for instance, we witnessed a polarised debate around genetic modification when activists destroyed a GM-potato trial field in 2011. Labelled by some as anti-scientific terrorism and by others as an act of civil disobedience in the name of democracy, this event has set off an acrimonious, divisive and deeply entrenched debate. Proponents as well as adversaries claim to have science on their side, invoking data about production yields, pesticide use, etc. Hereby, both sides often dismiss each other's point of view as being ideological, irrational or even criminal. Two events that took place in 2013 serve as striking examples of the polarisation: a public lecture of ecologist and anti-GMO activist Vandana Shiva and another one of GMO scientist Igor Potrykus. The latter passionately talked about his arduous efforts to develop golden rice, a GM rice variant that is claimed to pose a solution to the devastating problem of vitamin A deficiency in developing countries. When he concluded that GM opponents are responsible for the death of millions of children, qualifying NGO resistance against golden rice as ideological and criminal, he received a standing ovation from the audience. Tellingly, something very similar happened a few weeks earlier when Vandana Shiva, in her lecture entitled 'The Future of Our Food', passionately lamented against the monopolisation and commodification of food, thereby pointing out the crucial role of GMOs and their amenability to patenting. Here too, her conclusion that GMO-developing companies are criminal organisations was followed by a standing ovation.

Typology of sustainability problems

These illustrations show that some sustainability problems prove difficult to solve when there is (1) uncertainty and controversy regarding the knowledge base for solving them and (2) disagreement on the values and norms underlying alternative problem definitions and solution proposals. Problems such as climate change, loss of biodiversity, nuclear energy, resource depletion, poverty traps, etc. often have fuzzy boundaries and are usually linked to each other, and the many interactions between social and natural systems are of high and increasing complexity (Ostrom 2009). They are interpreted as so-called wicked problems (Rittel and Webber 1973) or unstructured problems (Hisschemöller and Hoppe 2001; Dijk et al. 2017) characterised by incomplete, uncertain or contested expert knowledge, conflicting values and objectives, a lack of unambiguous problem formulations and the impossibility to find uncontested definitive solutions. As such, these kinds of problems can only be understood in a context of complexity, uncertainty and diversity of values.[1] They contrast sharply with, for instance, the ozone hole problem described above.

Hisschemöller and Hoppe (2001) have made a typology of problems based on the degree of certainty of knowledge and the degree of agreement on values and norms. Figure 2.1 schematically visualises the varied types of problems, with the horizontal axis representing a continuum of increasing uncertainty regarding the available knowledge (uncertainty highest at the right end) and the vertical axis representing a continuum of (dis)agreement on norms and values (disagreement highest at the top end). Whereas the GMO controversy and climate change illustrate the characteristics of unstructured problems, ozone depletion can be seen as a typical example of a structured problem: clear and unambiguous scientific expertise regarding the problem and how to solve it was available and there was

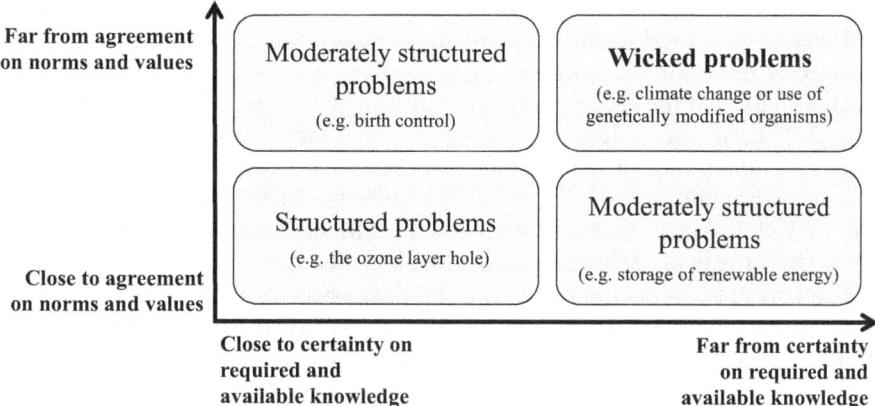

Figure 2.1 Typology of problems
Source: (based on Hisschemöller and Hoppe 2001)

little or no disagreement about which solutions were appropriate. An example of a moderately structured problem at the bottom-right end is the storage of electricity generated by renewable energy sources. There is a wide societal consensus about the desirability of this goal, yet the scientific expertise for designing sufficiently efficient and effective technologies to implement it is lacking. At the top-left end, an example of a moderately structured problem is birth control in view of reducing overpopulation and, thus, environmental impact. There is no lack of knowledge or technologies for implementing this, but the issue is very controversial because of the question of whether it is ethically acceptable. Many more examples of such problems can be found in the medical sphere and neuroscience. Scientific expertise and suitable technologies are largely available when it comes to, for instance, genetic and prenatal testing, the use of foetal tissue or the placement of certain neural implants. Yet, a lot of disagreement exists whether it is desirable, for instance, to enable women to give birth in their 50s, to eradicate Down syndrome or to use aborted foetal tissue in vaccines.

The call for post-normal science

Since the late 1980s, the idea that the production of scientific knowledge is not responding adequately to the challenges of our times and in particular those posed by the quest for sustainability has gained increasing acceptance with scientists and policymakers (Goeminne 2011). The confrontation with unstructured, wicked problems resulted in specific demands for science (Funtowicz and Ravetz 1993). Amongst others (Clark 2007; Kemp and Martens 2007; Ravetz 2018), Kates et al. (2001) advocate the development of a 'sustainability science' that considerably differs in structure, methods and content from science as we know it. It requires problem-driven and interdisciplinary research as well as participatory procedures that involve scientists, stakeholders, advocates, active citizens and users in the production of knowledge. Furthermore, sustainability science needs to be reflexive, i.e. sensitive to the way in which knowledge was generated and hence to what the underlying uncertainties are (Hulme and Toye 2006; Hugé et al. 2016). All this is deemed necessary to build more 'socially robust' knowledge on complex and contested issues (Nowotny et al. 2001). Scholars have proposed specific concepts to define forms of knowledge production that aim to take the 'wickedness' of problems seriously, such as Mode 2 science (Gibbons et al. 1994; Nowotny et al. 2003) and post-normal science.

This concept of post-normal science (PNS) has been introduced by Funtowicz and Ravetz (1992, 1993, 1994), who situated it next to two other problem-solving strategies in risk assessment: applied science and professional consultancy. The three strategies can be placed in a gradient on a diagram (see Figure 2.2) with the horizontal axis representing systems uncertainties, i.e. the technical, scientific and managerial complexities, and the vertical axis representing decision stakes, i.e. potential costs and benefits for the parties involved. When both system uncertainties and decision stakes are low, 'normal' science is useful: the available expertise is fully effective and applied science can continue its routine work. However, such 'puzzle-solving' (Kuhn 1962) normal science becomes inadequate when stakes and/or uncertainties rise to a medium level. Then, the skills, tacit knowledge, judgement and sometimes even courage of a practitioner are required and professional consultancy (e.g. by senior engineers) is more appropriate. But as we move further on the two axes to situations in which 'facts are uncertain, values in dispute, stakes high and decisions urgent' (Ravetz 1986, 422) – or, to

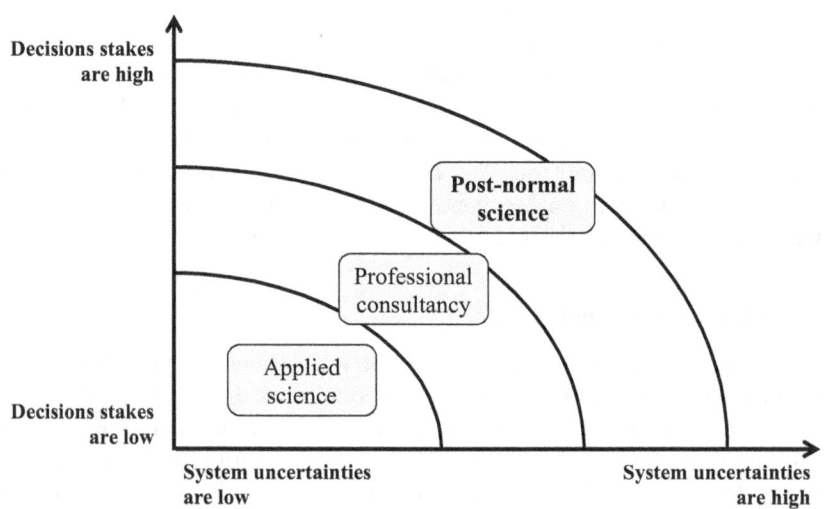

Figure 2.2 The post-normal science diagram

Source: (Functowicz and Ravetz 1993)

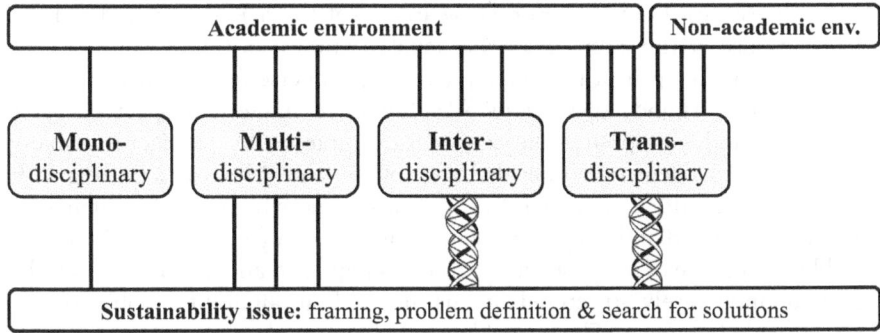

Figure 2.3 Mono-, multi-, inter- and transdisciplinary approaches

formulate it differently, when problems are wicked –this strategy also is seen as inadequate. Then, PNS is put forward, which implies reflexivity and the recognition of a multiplicity of legitimate perspectives on the issue at stake.

In PNS, transdisciplinarity is crucial. This is a form of knowledge production and problem solving that crosses disciplinary boundaries and involves cooperation between academics and other societal actors (e.g. policymakers, NGOs, companies, think tanks, etc.) in order to meet complex societal challenges (Klein et al. 2001). As such, this research strategy mobilises diverse knowledge, perspectives and expertise from inside and outside the academic world to investigate and frame problems and to look for solutions (see Figure 2.3). Monodisciplinary research draws on only one scientific discipline. A multi-disciplinary approach, in contrast, investigates an issue from the perspectives of several disciplines but without crossing the boundaries between these disciplines. The latter is different in interdisciplinary research. Here, there is cross-fertilisation between several disciplines. In transdisciplinary research, also non-academic actors, expertise and experiences are involved in knowledge production.

Towards post-normal education

Sustainability problems, especially those that are moderately structured or unstructured, do not only require a reconsideration of the mainstream forms of knowledge production but also pose major challenges to education. Teaching knowledge becomes challenging when there is discussion on scientific facts (Lönngren et al. 2016). When we cannot rely on an unambiguous scientific problem framing and clear-cut solutions for wicked sustainability issues, teaching and learning should move beyond a 'fact-based' approach of transferring clear knowledge and specific skills (see chapter 5). Dealing with the diversity and controversy that characterise unstructured problems also brings about ethical and political challenges for education. After all, what is at stake in these problems are questions regarding what are the right actions and good values and how to organise our society and a plurality of – sometimes conflicting – answers to these questions. Political controversy arises – at the macro societal level but sometimes also at the micro classroom level – when it becomes manifest that we are all caught up in a problem, yet through diverse and often conflictual concerns, interests and expectations. We often have different perspectives upon the problem and appropriate solutions for it (see chapter 1). The fact that many of those solutions demand a radical change of our way of life turns dealing with wicked sustainability problems into a political issue which involves deliberation (see chapters 8 and 9), dealing with a(nta)gonism (see chapter 9) and the possible emergence of political emotions (see chapters 8 and 20). Finally, teaching about moderately structured or unstructured sustainability problems also poses ethical challenges. The entanglement of facts and values in disputes over sustainability issues can easily bring about moral reactions and dilemmas that, depending on how a teacher handles these, can give rise to ethical reflection (see chapters 6 and 7).

Considering the important differences among varied types of sustainability problems (see the 'Typology of sustainability problems' section), sustainable development teaching should not be seen as one homogeneous kind of practice but, instead, as something that can take different forms depending on the context in which it takes place and the specificity of the content addressed. Hence, as teachers we should deliberately engage with the question of what kind of problems we are dealing with and, accordingly, choose a well-suited approach while designing our courses and lessons. Chapter 5 offers a useful framework in this regard by elaborating three distinctive teaching traditions. A 'fact-based' approach may be useful for teaching about structured sustainability problems. A 'normative' approach can be applied when there is little or no disagreement on norms and values and when the desirability of proposed solutions is uncontested. However, it is important to be aware of the possibility that controversy might always, unexpectedly, pop up – in the public debate as well as in classroom contexts. The more unstructured or wicked the sustainability problems we teach about, the more vital it becomes to create space for a 'pluralistic' approach.

Funtowicz and Ravetz's plea for post-normal science inspired us to introduce the notion of 'post-normal education' (PNE) (Block et al. 2018) in search of forms of teaching and learning that aim to take the challenges involved in dealing with wicked sustainability problems seriously. What does it imply to design teaching practices for dealing with problems where facts are uncertain, values in dispute, stakes high and decisions urgent? Importantly, it challenges educators to acknowledge both the wickedness and the urgency of sustainability problems. We are faced with a difficult tension between, on the one hand, a strong sense of urgency emerging from deep concerns about the severe implications of sustainability issues and, on the other hand, restraints against education as an instrument to foster predetermined ways of thinking and acting (Læssøe 2007; Wals 2010). Highlighting the uncertainty of knowledge and disputes over values captured in Funtowicz and Ravetz's (1993) dimension of 'systems uncertainties', researchers have criticised instrumental approaches that reduce education to a matter of learning to make proper choices among pre-existing alternatives based on universal factual or ethical guidelines (e.g. Ashley 2005; Östman 2010; Lundegård and Wickman 2007; Van Poeck and Vandenabeele 2012; Sund and Öhman 2014). Others, however, emphasise the 'decision stakes' dimension – stakes are high and decisions urgent – to argue for the importance of education as a means to break away from unsustainable lifestyles, behaviour, routines, worldviews and societal structures (e.g. Vare and Scott 2007; Jackson 2011; Kopnina 2012). Clear-cut recipes for successfully handling this tension, unfortunately, do not exist. As often in the teaching profession, it demands well-considered and context-specific (ad hoc) decision-making. However, some insights and principles can be useful to guide us in this challenging task (Block et al. 2018):

- *Treating 'facts' modestly.* As single right solutions for wicked sustainability problems do not exist, we need to leave room for uncertainties and a plurality of normative perspectives. Acknowledging the limitations of all knowledge-production systems and the socially embedded, situated character of the genesis of scientific knowledge (Latour 1986; Jasanoff 2004; Nowotny et al. 2005) requires a modest attitude towards facts claimed by experts. Or, as further elaborated in chapter 3, 'pure' facts alone cannot guide our actions towards a sustainable world. We will always also need value judgements to make choices of which facts we believe to be most relevant, fruitful, useful, efficient, etc. in relation to the concrete sustainability problem we are trying to solve. This implies recognising, in our teaching practice, that every scientific construction includes some concerns while excluding others (Goeminne and François 2010). Both in the production of knowledge through scientific practice and in its dissemination through education practice, the framing of sustainability issues is inevitably selective, putting some elements at the forefront and leaving others out of view. Roberts and Östman's (1998) empirical investigations of science education textbooks, for instance, reveal how the presented knowledge is always entangled with specific views of nature and how to treat it. This offers students very particular strategies (e.g. technology-driven, market-driven, government-driven, grassroots-driven, sufficiency-driven . . .) for reasoning about the world and their place in it while omitting others. As such, the learning of knowledge is always accompanied by the learning of values or so-called companion meanings (Östman 1996) (see chapter 4).
- *Engaging, pluralistically, with the basic principles of sustainable development.* As argued in chapter 1, sustainable development is a concept that can easily be rendered into a consensual catch-all term that only serves to safeguard the status quo. Taking its wickedness seriously implies recognising that there are no simple definitions or objective indicators that allow us to unambiguously define what is (un)sustainable once and for all. Taking responsibility for the urgency of sustainability problems, however, requires a sustained effort to draw attention to the basic principles and characteristics of sustainable development, i.e. that it is a form of development that is socially just, remains

within the carrying capacity of the earth's ecosystems, guarantees quality of life for all and realises that within a democratic society. Although notions such as justice, ecological limits, quality of life and democracy can also be seen as essentially contested concepts that give rise to struggles over different interpretations, they do offer us touchstones to engage in such a pluralistic exploration and dialogue and to avoid 'anything-goes' approaches – for instance, by trying to go beyond the dominant 'weak' approaches of sustainable development (see chapter 1) in our research and teaching activities (Block et al. 2018).

• *Designing issue-driven and problem-oriented teaching and learning practices.* Rather than in abstract discussions about sustainable development in general, interesting opportunities to learn about uncertain facts, disputed values, high stakes and urgently needed decisions emerge when students are exposed to concrete sustainability problems. Such an issue-driven and problem-oriented approach creates time and space for students to study a sustainability problem, its consequences and possible solutions. In so doing, they will discover a multiplicity of knowledge claims, concerns, interests, values and norms entangled in it. Teachers can thereby reveal their own perspectives, concerns, value judgements, attachments, etc., thus stimulating students to explore it further, to confront it with other perspectives and, finally, to relate to it by formulating their own, singular perspective, concern and judgement. Such inquiry (see chapter 10) – preferably set up in a transdisciplinary way (see the previous section, 'The call for post-normal science') – can lead to well-informed judgement based on careful observation, a wide range of information and acquaintance with a plurality of experiences, skills, knowledge and perspectives (Dewey 1938/2015).

Acknowledgements

The research on which this chapter is based on was supported by Ghent University (International Thematic Network 'SEDwise – Sustainability Education: Teaching and learning in the face of wicked socio-ecological problems') and by Research Foundation Flanders (Scientific Research Network 'Public pedagogy and sustainability challenges').

Note

1 This argument should not be understood as an 'anything-goes', relativist position towards scientific evidence regarding issues like climate change. We are not arguing for treating all possible 'alternative facts' as equally valid. On the contrary, it would be very dangerous if sustainable development teaching would contribute to the so-called post-truth era, in which scientific findings are treated as mere matters of faith and public tolerance of inaccurate, undefended allegations and outright denials of facts reaches worrying levels (Higgins 2016). Yet, as we further elaborate in chapter 3 and chapter 5, scientific facts can neither function as the one and only, objective basis for so-called value-free education providing students with all the necessary knowledge for solving

sustainability problems. Facts and values are intertwined in sustainability issues, for instance because we need values to judge which facts matter more than others (see chapter 3).

References

Ashley, M. (2005) "Tensions between indoctrination and the development of judgement: The case against early closure". *Environmental Education Research*, 11(2): 187–197.

Block, T., Goeminne, G. and Van Poeck, K. (2018) "Balancing the urgency and wickedness of sustainability challenges: Three maxims for post-normal education". *Environmental Education Research*, 24(9): 1424–1439.

Clark, W.C. (2007) "Sustainability science: A room of its own". *Proceedings of the National Academy of Sciences of the United States of America*, 104: 1737–1738.

Dewey, J. (1938/2015) *Experience and Education*. Free Press, New York.

Dijk, M., de Kraker, J., van Zeijl-Rozema, A., van Lente, H., Beumer, C., Beemsterboer, S. and Valkering, P. (2017) "Sustainability assessment as problem structuring: Three typical ways". *Sustainability Science*, 12(2): 305–317.

Farman, J.C., Gardiner, B.G. and Shanklin, J.D. (1985) *Nature*, 315: 207–210.

Funtowicz, S.O. and Ravetz, J.R. (1992) "Three types of risk assessment and the emergence of Post-Normal Science". In Krimsky, S., Golding, D. eds., *Social Theories of Risk*. Greenwood, Westport, 251–273.

Funtowicz, S.O. and Ravetz, J.R. (1993) "Science for the post-normal age". *Futures*, 2(7): 739–755.

Funtowicz, S.O. and Ravetz, J.R. (1994) "Uncertainty, complexity and post-normal science". *Environmental Toxicology and Chemistry*, 13(12): 1881–1885.

Gibbons, M., Limoges, C., Nowotny, H. et al. (1994) *The New Production of Knowledge*. Sage, London.

Goeminne, G. (2010) "Climate policy is dead, long live climate politics!" *Ethics, Place and Environment*, 13(2): 207–214.

Goeminne, G. (2011) "Has science ever been normal? On the need and impossibility of a sustainability science". *Futures*, 43(6): 627–636.

Goeminne, G. and François, K. (2010) "The thing called environment: What it is and how to be concerned with it". *The Oxford Literary Review*, 32(1): 109–130.

Grundmann, R. (2016) "Climate change as a wicked social problem". *Nature Geoscience*, 9: 562–563.

Higgins, K. (2016) "Post-truth: A guide for the perplexed". *Nature*, 540(9), pre-published online: https://doi.org/10.1038/540009a

Hisschemöller, M. and Hoppe, R. (2001) "Coping with intractable controversies: The case for problem structuring in policy design and analysis". In Hoppe, R., Hisschemoller, M., Dunn, W.N., Ravetz, J.R. eds., *Knowledge, Power and Participation in Environmental Policy Analysis*. Transaction Publishers, New Brunswick, London, 47–72.

Hugé, J., Block, T., Waas, T., Wright, T. and Dahdouh-Guebas, F. (2016) "How to walk the talk? Developing actions for sustainability in academic research". *Journal of Cleaner Production*, 137: 83–92.

Hulme, M. (2009) *Why We Disagree About Climate Change: Understanding Controversy, Inaction And Opportunity*. Cambridge University Press, Cambridge.

Hulme, D., Toye, J. (2006) "The case for cross-disciplinary social science research on poverty, inequality and well-being". *Journal of Development Studies*, 42: 1085–1107.

Jackson, M.G. (2011) "The real challenge of ESD". *Journal of Education for Sustainable Development*, 5(1): 27–37.

Jasanoff, S. ed. (2004) *States of Knowledge: The Co-Production of Science and Social Order*. Routledge, London and New York.

Kates, R.W., Clark, W.C., Corell, R. et al. (2001) "Sustainability science". *Science*, 292: 641–642.

Kemp, R. and Martens, P. (2007) "Sustainable development: How to manage something that is subjective and never can be achieved". *Sustainability: Science, Practice and Policy*, 3(2): 5–14.

Klein, J.T., Grossenbacher-Mansuy, W., Häberli, R., Bill, A., Scholz, R.W. and Welti, M. eds. (2001) *Transdisciplinarity: Joint Problem Solving Among Science, Technology, and Society. An Effective Way for Managing Complexity*. Birkhauser, Basel.

Kopnina, H. (2012) "Education for sustainable development (ESD): The turnaway from 'environment' in environmental education?" *Environmental Education Research*, 18(5): 699–717.

Kuhn, T.S. (1962) *The Structure of Scientific Revolutions*. University of Chicago Press, Chicago.

Læssøe, J. (2007) "Participation and sustainable development: The post-ecologist transformation of citizen involvement in Denmark". *Environmental Politics*, 16(2): 231–250.

Latour, B. (1986) *Science in Action*. Open University Press, New York.

Lönngren, J., Ingerman, Å. and Svanström, M. (2016) "Avoid, control, succumb, or balance: Engineering students' approaches to a wicked sustainability problem". *Research in Science Education*, 47(4): 805–831.

Lundegård, I. and Wickman, P.-O. (2007) "Conflicts of interest: An indispensable element of education for sustainability". *Environmental Education Research*, 13(1): 1–15.

Nowotny, H., Scott, P. and Gibbons, M. (2001) *Re-thinking Science-Knowledge and the Public in an Age of Uncertainty*. Polity Press, Cambridge.

Nowotny, H., Scott, P. and Gibbons, M. (2003) "Introduction: 'Mode 2 revisited': The new production of knowledge". *Minerva*, 41: 179–194.

Nowotny, H., Scott, P. and Gibbons, M. (2005) "Rethinking science: Mode 2 in societal context". In Carayannis, E.G. and Campbell, D.F.G. eds., *Knowledge Creation, Diffusion, and Use in Innovation Networks and Knowledge Clusters: A Comparative Systems Approach Across the Unites States, Europe and Asia*. Praeger, Westport.

Östman, L. (1996) "Discourses, discursive meanings and socialisation in chemistry education". *Journal of Curriculum Studies*, 28(1): 37–55.

Östman, L. (2010) "Education for sustainable development and normativity: A transactional analysis of moral meaning-making and companion meanings in classroom communication". *Environmental Education Research*, 16(1): 75–93.

Ostrom, E. (2009) "A general framework for assessing the sustainability of socioecological systems". *Science*, 325: 419–422.

Ravetz, J.R. (1986) "Usable knowledge, usable ignorance: Incomplete science with policy implications". In Clark, W.C. and Munn, R.C. eds., *Sustainable Development of the Biosphere*. Cambridge, New York, 415–432.

Ravetz, J.R. (2018) "Post-script: Heuristics for sustainability science". In König, A. ed., *Sustainability Science: Key Issues*. Routledge, Abingdon.

Rittel, H. and Webber, M. (1973) "Dilemmas in a general theory of planning". *Policy Sciences*, 4: 155–169.

Roberts, D. and Östman, L. eds. (1998) *Problems of Meaning in Science Curriculum*. Teachers College Press, London.

Stolarski, R.S., Krueger, A.J. et al. (1986) "Nimbus 7 satellite measurements of the spring-time Antarctic ozone decrease". *Nature*, 322: 808–811.

Sund, L. and Öhman, J. (2014) "On the need to repoliticise environmental and sustainability education: Rethinking the postpolitical consensus". *Environmental Education Research*, 20(5): 639–659.

Van Poeck, K. and Vandenabeele, J. (2012) "Learning from sustainable development: Education in the light of public issues". *Environmental Education Research*, 18(4): 541–552.

Vare, P. and Scott, W. (2007) "Learning for a change: Exploring the relationship between education and sustainable development". *Journal of Education for Sustainable Development*, 1(2): 191–198.

Wals, A.E.J. (2010) "Between knowing what is right and knowing that is it wrong to tell others what is right: On relativism, uncertainty and democracy in E(E)SD". *Environmental Education Research*, 16(1): 143–151.

3 Principles for sustainable development teaching

Leif Östman, Katrien Van Poeck and Johan Öhman

Introduction

We witness repeated appeals to education for contributing to building a more sustainable world. However, as the first two chapters show, the specificity of sustainability issues also offers specific challenges for teaching and learning. These are elaborated below in terms of the need to move beyond traditional 'schooling' practices, to avoid both instrumentalism and relativism, and to be aware that sustainable development teaching and learning includes ethical and political dimensions. Furthermore, we shortly introduce the approach to teaching and learning that is underlying the ideas, frameworks and models presented in this book. In chapters 10 and 11, we elaborate further on the approach.

Teaching and learning challenges

As described in chapter 2, there are many situations where we aim to create more sustainable living, but where a stable and uncontested knowledge base is missing and where people have different values and opinions regarding, for example, suitable solutions to major sustainability problems. Teachers and schools also face such issues, both in the classroom and when it comes to striving for a sustainable campus. We have, for example, worked with schools and teacher teams that aimed to involve students in making the campus's waste management more sustainable. In the classrooms, they taught about 'Lansink's Ladder', a hierarchy in waste management recommending reuse over recycling. In line with that, they used reusable bottles on the campus and some schools forbid the students to bring recyclable drink cartons. Repeatedly, this brought about controversy – both factual and normative. Some parents and students contested the fact that reusing glass bottles has a smaller ecological impact than recycling drink cartons due to the cleaning and transportation of the bottles, thereby referring to results of scientific studies published in newspapers. The schools tried to solve this by getting expert advice, but it turned out that a scientific consensus on this issue did not exist – at least at that moment. Experts lacked sufficient data to come up with unambiguous results based on 'life cycle analyses' of all products. Another kind of controversy that arose was not focused on facts but rather revealed disagreement on values and norms. Can a school impose certain pro-environmental behaviour on the students? Is it desirable to force students to prioritise care for the environment over freedom of choice? For example, what if a student wants to bring soy milk (only available in drink cartons) because she is allergic to the cow milk in the glass bottles at school? In all these situations – 'unstructured problems' as we called them in chapter 2 – it is obvious that a transition to a more sustainable solution forces participants to handle the different moral and political standpoints that they bring to the table. Thus, when faced with sustainability issues that are characterised as unstructured problems, participants need to solve these issues as they go along in the work towards a transition. This requires genuine creativity: they need to learn-by-doing as they do-by-learning. In such a complex process, one obviously needs a lot of basic knowledge and skills – and creativity and commitment. Even if the problems are structured, we need to make judgements regarding which knowledge we should prioritise in decisions about solutions, because as soon as we bring in three perspectives on an issue – the ecological, the economic and the social (see chapter 1) – the amount of knowledge becomes so great that one decision cannot possibly accommodate all of it. Thus, we have to prioritise and involve values and norms. If we want our students to be properly equipped for understanding (wicked and structured) sustainability problems and be constructive partners in looking for solutions, we need to design environmental and sustainability education (ESE) that goes beyond traditional, instrumental 'schooling' practices.

Principles for sustainable development teaching

As we will elaborate below by sketching five interrelated principles, ESE is a matter of creating engagement and commitment in relation to the contents of teaching – for example changes in lifestyles in order to handle climate change – instead of just focusing on the evaluation system of education (getting good grades). It also means that in dealing with the subject knowledge in the classroom, we should be attentive to how students can apply it to other contexts (e.g. looking for sustainable solutions) instead of assuming that they will automatically do that. Furthermore, we need to find ways to handle facts, values and norms in a pluralistic way and acknowledge the ethical and political dimension of the content as an integral part of teaching and learning.

Principle 1: *create engagement for the content of teaching*

The first principle for designing ESE that goes beyond education as instrumental schooling is to encourage students to become engaged. An important part of students' engagement is that they become emotionally attached to the content that is dealt with. Emotional attachments grow and take shape in the process of learning, because students generally lack a deep experience of much of the content they encounter in schools. However, research has shown (e.g. Ojala 2012, 2016) that many students experience profound feelings of anxiety, worry and helplessness but also hope regarding sustainability issues such as climate change. Addressing these issues and related feelings can substantially increase the motivation to learn subject knowledge but also create engagement in relation to the content and, hence, make the often-advocated move from surface learning to deep learning. Often, Ojala (2016) argues, worry about societal problems is perceived as something only negative, but it can also motivate students to think critically, to deliberate, to engage in problem solving and experimentation or to break with unsustainable habits and practices. In other words, it can engage students in an 'inquiry' (see below).

The American philosopher and educationalist John Dewey (1934) introduced the concept of '*consummatory experience*' to highlight the emotional attachments that are connected to the process of inquiry. It concerns the experience of fulfilment of, for example, a problem-solving activity. Not all our experiences of thinking and/or doing lead to consummatory experiences, as some are

> too automatic to permit a sense of what it is about and where it is going. It comes to an end but not to a close or consummation in consciousness. Obstacles are overcome by shrewd skill, but they do not feed experience. There are also those who are wavering in action, uncertain, and inconclusive like the shades in classic literature. Between the poles of aimlessness and mechanical efficiency, there lie those courses of action in which through successive deeds there runs a sense of growing meaning conserved and accumulating toward an end that is felt as accomplishment of a process.
>
> (Dewey 1934, 40)

When we succeed with problem solving in such a way, we are struck by a positive bodily feeling, which often becomes immediately expressed in spontaneous aesthetic utterances as 'cool', 'nice' and 'beautiful' (Wickman 2006; Jakobsson and Wickman 2008). Such feeling in relation to the activity and the content dealt with is a crucial aspect of creating interest and even commitment. Thus, it is of great importance that teaching makes it possible for students to have consummatory experiences, which requires that the staged inquiry (see below) takes departure in the students' earlier experiences. But equally important is that the problem that is the cause of the inquiry is challenging enough for the students. If the problem is too easy, the inquiry becomes inauthentic for the students and the experience of fulfilment will not happen, because the problem is not a problem for them.

Principle 2: use the right focus for the teaching

The learning objectives in the curriculum, national tests, handbooks, etc. continuously remind teachers of the abundance of subject knowledge we have to teach. At the same time, ESE aims to be societally relevant. That is, we want to help the students to contribute to creating a more sustainable world. Thus, our ambition is that students are able to apply subject knowledge from science, mathematics, economics, civics, geography, art, craft, etc. to everyday life and societal problems. In the history of teaching, and also today, one can find different views and practices as to how teaching should be organised in order to achieve this, namely 'induction into a discipline' and 'learning from a discipline'. With discipline, we here mean a branch of knowledge like the disciplines at universities but also arts and crafts. These two approaches in teaching represent two different *subject focuses* (Östman 1995, 1996, 1998; Chambers 2008). The background of the concept of subject focus is that it is amazing how much is said and done during a lesson, and we cannot expect the students to learn all of it. But we do want them to pay attention to and learn the essentials. Thus, the lesson always focuses on something; it has a primary object. We can say that there is always a subject focus.

In an induction into a discipline focus, the purpose is that students shall be inducted into a specific discipline such as science, mathematics, economics, geography, art, craft, etc. The ambition is that all students should become a scientist, mathematician, economist, geographer, artist, crafter, etc. if not to a hundred percent, at least a mini version of it. In this 'fact-based' approach to ESE (see chapter 5), the assumption is that students will *automatically* be able to apply the subject knowledge to everyday life and societal problems. The only thing that education needs to be concerned about is to facilitate students' learning of basic knowledge and skills: if they learn that, they will be able – without any further training – to apply them to any everyday or societal problem of relevance.

In a learning from a discipline focus, the purpose is that specific knowledge and skills from a discipline is used by the students in relation to everyday issues or problems that require political and ethical reflections, argumentations and decisions. Here the students will learn the basic knowledge and skills while they apply it. Hence, students will learn the basic knowledge and skills at the same time as they practice how to apply them to everyday or societal problems. In the learning from a discipline focus, one can find two categories: *learning from a discipline for everyday situations* and *learning from a discipline for political and ethical situations* (see Table 3.1). The main difference between those two categories is that the first concerns the application of knowledge and skills from disciplines onto technical solutions, like for example how to sort garbage into different categories relevant for the recycling process. In the second category, the knowledge and skills are applied to an issue where a prioritisation between different alternatives has to be made, which means that value-laden judgements need to be executed, as in the example above where a controversy arose between two different alternatives to sustainable consumption: recycling or reuse. These two different categories of learning from a

Table 3.1 Different subject focuses

1. Induction into disciplines	Students are learning sustainability knowledge and skills of disciplines
2. Learning from disciplines:	Students are learning sustainability knowledge and skills of disciplines and how to apply them on different types of sustainability problems:
a) for everyday situation	technical problems
b) for political and ethical situations	ethical and political problems

discipline focus thus involve different types of application skills, and therefore it is important to distinguish them from each other (see further Östman 1998).

In an ESE perspective, the learning from a discipline focus is preferred because in this teaching approach students not only learn, for example, scientific knowledge and skills but also how to apply them in their lives, now and in the future. Students are rarely by themselves able to figure out how to concretely apply, for example, ecological knowledge into their life as consumer or citizen. This is no surprise, since this is a complex transformation and research has also shown that there is no straightforward connection between knowledge and behaviour change (Kollmuss and Agyeman 2002; Boeve-de Pauw and Van Petegem 2018). In this transformation, elements such as values and motivation are found to play a crucial role. Thus, the learning from a discipline focus is an approach that needs to be used in combination with principle 1: to create engagement. Another argument for using it is that when students work with applying, for example, a scientific model to an everyday problem such as a sustainability issue, they will also naturally deepen their understanding of the model. As such, the two goals of teaching all the subject knowledge in the curriculum and educating students who are able to contribute to a more sustainable world can go hand-in-hand.

Principle 3: deal with local sustainability problems

Addressing authentic local environmental and sustainability issues with direct connections to students' life situation is one way of increasing the relevance of the teaching content, and hence students' motivation and engagement. Thus, a learning from a discipline focus with emphasis on a local sustainability problem is highly relevant in ESE. By addressing local issues, students will get the chance to become change agents, i.e. they can be directly involved in solving issues that the local community is struggling with. The focus on the local does not take away the importance to connect the local with the global: the point is that the students will be able to get a first-hand experience of how dealing with sustainability problems involves ecological, social and economic aspects. Such a first-hand experience is highly crucial because it gives the students practice in how to take into account these three different perspectives and use and apply knowledge from science, mathematics, economics, civics, geography, art, craft, etc. in the problem-solving process (see principle 2 and the description of 'learning from a discipline focus'). In order to support students' work to handle ecological, social and economic perspectives, teachers from different subjects need to cooperate. Although working with authentic local sustainability problems can require that we need to change our habits of teaching, the results can be very positive when it comes to students' learning (see for example Östman et al. 2013). One model that facilitates the planning of teaching and the cooperation between teachers of different subject is LORET (locally relevant teaching). It offers teacher teams a four-phase step-by-step procedure to plan locally relevant sustainable development teaching that is adapted to local needs and conditions while, at the same time, allows teaching subject knowledge in such a way that we realise the objectives in the curriculum (see Textbox 3.1).

Textbox 3.1   LORET: locally relevant teaching

Phase 1. Identify key sustainable development issues in your local community

Steps:

1   Gather a number of colleagues, preferably from different subjects.
2   Each person lists key issues for sustainable development in the local community (individual work).
3   Discuss the different key issues in the group and choose one to work with in developing one LORET (the excluded key issues can be used in other LORETs).

Phase 2. Identify goals for sustainable development

Steps:

1   Individual work: List the goals for the sustainable development you want to achieve in the local community. Ask yourself what would need to happen in order to achieve them.
2   Discuss in the group and make a common list.

Phase 3. Identify the knowledge needed to reach the identified goals

Steps:

1   Individual work: Wise actions require knowledge. Make a list of the things students need to know about in order to reach the goals. This can include scientific knowledge and practical skills but also, for instance, ethical and/or democratic competences.
2   Compare the lists in the group.
3   Make a mind map: Construct headlines for the goals and sort the necessary knowledge under the headlines.
4   Determine the school subjects that will be needed in order to cover the knowledge areas marked in your mind map. Indicate these on your mind map using different colours for different subjects. You will probably find that several school subjects are necessary.

Phase 4. Creating a teaching plan: LORET

The mind map is a perfect basis for creating a teaching plan for locally relevant sustainable development teaching. The key issue you have chosen is the teaching theme. The goals in the mind map can be regarded as topics for one lesson or a series of lessons. Thus, a theme can consist of several topics and each topic can embrace one or several lessons.

Steps:

1   In the mind map, you have listed things that students need to know about for each goal. You have also highlighted which school

subjects need to be involved in order to provide the students with the necessary knowledge. For each school subject, list the subject knowledge that you want to teach the students.

2   If more than one school subject is involved, decide how you will organise the work so that the students can effectively integrate the knowledge from the different school subjects.

3   Identify connections to the local community: Do the students need to gather knowledge in the local community, for example by interviewing a farmer, a parent, etc.? Will students take any action in the local environment on the school estate or in the schoolyard? In the local community? At home?

4   Determine the number of lessons that will be needed and the content for each.

5   Identify teaching methods that will help students to get involved and learn the required knowledge.

6   Identify connections to the curriculum. This step is crucial, because it justifies sustainable development teaching. In step 1, you identified which school subject knowledge would be taught to the students. It is very likely that your identified knowledge is already prescribed in the syllabi: refer to the page in the syllabus or use a quotation.

7   Critically examine the plan and revise it as necessary: Is this a key issue for development in the region? Are ecological as well as economic and social aspects covered? Does the plan include argumentation- and decision-making skills? Are there connections to the local society? Does the plan build upon integration between different subjects? What action components are involved? Will students learn both practical and theoretical SD knowledge? etc.

8   Write the final plan. We strongly recommend that you document the LORETs in such a way that they can be read and used by other teachers and, as such, start to function as local curricula.

Principle 4: stress pluralism

The fourth principle for ESE is *pluralism*, that is, to bring in many different perspectives – economic, ecological and social, as well as local and global – and different types of knowledge – e.g. academic, indigenous, grassroots, lay knowledge, etc. – in order to achieve wise prioritisation and decision-making. This principle is applicable regardless if the sustainability issues are wicked or not. The main argument for pluralism is that the involvement of three different perspectives (the social, the ecological and the economic) on sustainable development will force people to make prioritisations (see above). Thus, highlighting pluralism in view of transitions to sustainability can simply be described in this way: many people's thinking and valuing is generally more productive than if just one person does that work. Another reason is that a pluralistic approach brings about a possibility for many people to become involved in decision-making, which could be seen as an important element in a democratic way of organising a society. One of the main doubts raised in relation to pluralism is that it can invoke naive relativism. If all utterances, suggestions, etc. are regarded as of equal value in decision-making, we do not make a difference between knowledge on the one hand and subjective values on the other which, according to the criticism, often results in the attitude of 'anything goes'. Another possible consequence is that 'alternative facts' are all there is. Such positions are risky because it can give rise to the attitude that nothing matters, which takes away the incentives for people to take part in decision-making. Some defenders of pluralism argue that this criticism is based on the belief that the search for universal truth is the only solution, because such truth will make all decision-making rational, i.e. it eradicates the need to take values into account. Other defenders of pluralism, however, argue that universal truth does not exist since it is humans who create knowledge. Because of this, knowledge will always be value-laden.

This discussion seems to offer only two alternatives, but there is actually a third perspective, which William James (1842–1910) – an early pragmatist and one of the fathers of American psychology – presented already in the beginning of 1900 (see also Östman 2010; Van Poeck 2018). James's point is that all our actions that are preceded by thinking involve values, but this does not make knowledge to be something subjective. His argument is as follows:

> conceive yourself, if possible, suddenly stripped of all the emotion with which your world now inspires you, and try to imagine it as it exists, purely by itself, without your favourable or unfavourable, hopeful or apprehensive comment. It will be almost impossible for you to realize such a condition of negativity and deadness. No one portion of the universe would then have importance beyond another; and the whole collection of its things and series of its events would be without significance, character, expression, or perspective. Whatever of value, interest, or meaning our respective worlds may appear endued with are thus pure gifts of the spectator's mind.
>
> (James 1905/1985, 177)

Imagine that you have access to all facts in the world, but all your values have disappeared. You live in a world where the only things that exist are true facts. You enter the shop – what will you buy for making a sustainable dinner tonight? The problem you will encounter is that there exist so many facts regarding sustainable food that you cannot consider them all when shopping for just one dinner. One fact might be, for instance, that fair trade products improve the living conditions of farmers in the Global South. Another fact is that shipping rice – fair trade or not – from Asia to Europe requires energy and thus has an ecological impact. You need to choose which true facts will guide your shopping. But here you will face a problem: to choose requires judgement, which in turn requires some form of value. Since all values are gone, a choice will not be possible. Thus, James's punch line is that in a world consisting solely of facts, we will not be able to act. We need values to make things happen!

This pragmatist perspective does acknowledge the importance of values and value judgements in human actions (e.g. decision-making), but without reducing knowledge to something subjective. What is proposed is that both knowledge and value are necessary for making actions and activities possible. Paraphrasing the philosopher Ludwig Wittgenstein (1889–1951), one could say that knowledge functions as hinges in a door (Wittgenstein 1969, §655). Knowledge makes it possible to act intelligently, as the hinges enable the door to move. But important to notice is that in the future we might create new knowledge that we want to use as hinges for our activities. Thus, one could see knowledge as temporary resting places (Rorty 1982) for making it possible to act intelligently. But knowledge is not enough: we need to make value judgements (choices) of which knowledge we want to become the hinges.

Taking departure in the above reasoning, we stress the importance of a pluralistic approach to sustainable development teaching (see chapter 5), i.e. forms of teaching that illustrate that with knowledge we have the pre-conditions for action, but that we also need value judgements to make choices of which knowledge we believe to be most relevant, fruitful, useful, efficient, etc. in relation to the concrete problem of sustainability we are trying to solve. By stressing pluralism in this way, we avoid an anything-goes attitude and, at the same time, give incentives for students to get involved in decision-making regarding sustainability problems.

Principle 5: include ethical and political dimensions

The fifth principle is to encourage the students to become knowledgeable and skilled in dealing with the political and the ethical dimensions of sustainability issues. The reason for this is what we emphasised above, namely that the application of three different perspectives – the social, the ecological and the economic – on sustainability issues forces people to make prioritisations. Such prioritisation involves values. Some of these values are connected to the ethical and political dimensions of an issue, but not all values are. In order to be able to identify those values that are connected to ethical and political aspects of sustainable issues, Wittgenstein's (1993) distinction between 'relative values' and 'absolute values' serves us well.

Relative values are important for making choices of which knowledge we believe to be most relevant, fruitful, useful or efficient. Relative values are connected to the purposes of a particular activity. If we, for example, say that something is an efficient solution for energy saving in a house, we refer to a fixed scale and criteria that are connected to this specific activity: this particular solution reduces more watt usage than do other solutions, and thereby we judge it as being efficient. Efficient is here an example of a relative value.

When we communicate the ethical and political aspect of sustainability issues and situations, we generally do so by expressing *absolute values* (see chapter 6). We then express that something has a universal value, a value that is not connected to a specific activity. If we, for example, say that it is wrong to cause unnecessary harm to animals, we communicate that this 'wrong' is universally valid, not conditioned to specific circumstances. This communication makes it possible for other people to understand that we are making an ethical judgement. This does not necessarily mean that we believe that the value judgement is valid for everyone everywhere, but rather that it *ought* to be a general concern.

The difference between the political and the ethical is mainly related to the object for our absolute value judgements. Simplified, ethical situations and issues concern the right and correct way of acting towards humans and nature, while a political concern the best and correct way of organising the society. In everyday situations, the political and ethical are sometimes entangled. For instance, ethical concerns of how people should be treated are entangled with concerns on how to organise society in discussions on labour regulations. This entanglement becomes visible when people argue that it is wrong to exploit people, and therefore there should be good laws for the organisation of labour relations that prevent employers from exploiting their employees.

Transactional theory on teaching and learning

Earlier, we described five principles that are important to take into consideration when planning and designing sustainability teaching. In the following, we will supplement those principles by elaborating on what to consider for making the planned teaching successful, namely that the students will actually learn what we have designed for. We will do so by presenting a transactional theory of teaching and learning (which will be further elaborated in chapters 10 and 11) after briefly outlining two starting points for our way of approaching learning. The first starting point is that we use the term learning in relation to three overall purposes of teaching, namely qualification (knowledge, skills), socialisation (values, worldviews) and person-formation. These three broad purposes are elaborated in chapter 4.

The second starting point is the so-called didactic triangle (see Figure 3.1). The didactic triangle has to be seen as situated in an institutional context of education, which means that the three earlier mentioned overall purposes are governing participants' actions, as they give an overarching direction for what they are supposed to achieve in the activities. Furthermore, there are at least two different participants involved in the activities, namely a student and a teacher. In order to achieve these overall goals, there must be something to be taught and learned, i.e. there needs to be content involved in the activities.

Most learning theories do perceive the interplay between humans and the surrounding world as the engine for learning. We are from birth, and probably even before, affected not only by the social world but also by the physical. This is one of the reasons that Dewey and Bentley (1949/1991) chose to introduce the term *transaction*. The transactional perspective emphasises that humans are constantly trying – through action – to coordinate with the surrounding social and physical world. The coordination has a function: either to achieve something we planned for or to adapt to changes in the surrounding world. Because of this coordination,

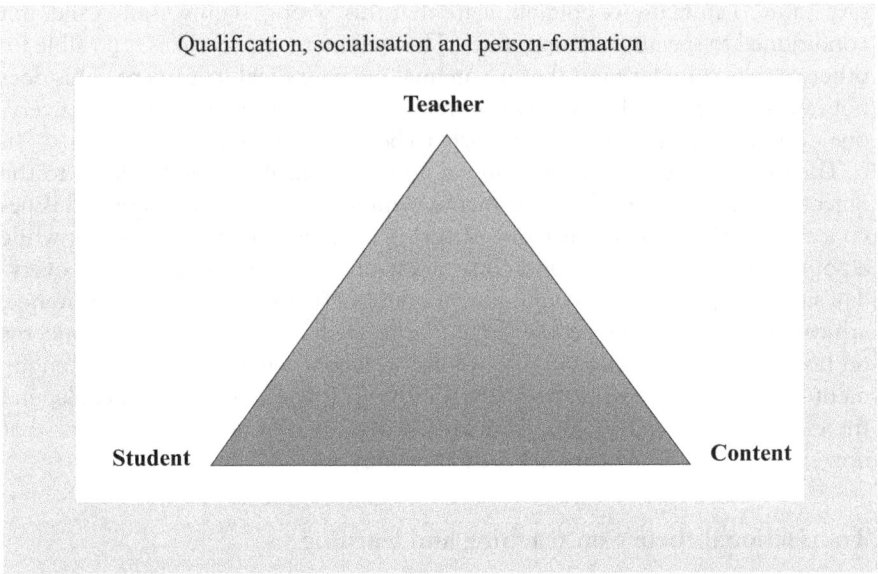

Figure 3.1 The figure shows the 'didactic triangle' in an institutional setting
Source: Author's own.

we learn. Learning can thus be described in terms of action, i.e. 'as meaning making resulting in a more developed and specific repertoire for coordinating activities with the environment' (Östman and Öhman 2010, 5). What drives learning is a continuous process of doing and undergoing the consequences of acts (Garrison 2001). Dewey (1938, 35) understands education as a process that takes place through encounters between a person and the surrounding world:

> every experience enacted and undergone modifies the one who acts and undergoes, while this modification affects, whether we wish it or not, the quality of subsequent experiences. For it is a somewhat different person who enters into them . . . . every experience both takes up something from those which have gone before and modifies in some way the quality of those which come after.

When we teach, we change students' surrounding world by directing their attention to specific objects. The objects can be cultural – for example, words, sentences, a figure, a map – or physical objects, i.e. a stone, a leaf. When we direct students' attention, we try to set a learning environment for the students: we set the scene. Usually we also ask them to *do* something with that learning environment: to act on the scene (see further chapter 11). We can, for instance, set the scene by having the students collect water animals for a biological study of water quality and then let them determine the quality of the water by identifying the collected animals. Such self-activity of the students can take the form of an *inquiry* (see Figure 3.3). Both of these interventions – to set a learning environment and to stage a self-activity of the students – are the central tools we use as teachers to facilitate students' learning.

Constructivist learning research, which often takes departure in the work of the influential Swiss researcher Jean Piaget (1896–1980), has shown that what a student will learn in the activity depends on her/his earlier experiences (knowledge, skills and values) (Piaget 1989). It is with the help of previous experiences that students are trying to make the new objects they encounter intelligible. Since each student brings unique experience into the learning activity, it means that in a class of 20 students, the result will be 20 somewhat different learning outcomes. Sociocultural research on learning often takes inspiration from the Russian researcher Lev Vygotsky (1896–1934), who became very well known for his research on the influence of the sociocultural environment on people's learning (Vygotsky 1978). In sociocultural research, it is explicated how the teacher, peers and cultural artefacts (objects) influence the learning of individual students. Thus, artefacts such as a map function as a tool, but the tool is not a neutral tool. It also influences how we experience things and, hence, how we learn. In chapters 10 and 11, we exemplify and elaborate on this in more detail.

In transactional research on learning, both the current situation and students' earlier experiences are considered. The result of such research is particularly useful for us as teachers, since it is crucial to stage the learning environment and design the activity in such a way that students can use their old experiences in a fruitful way. That is, to make it possible for the students to understand the offered learning environment and activity and *at the same time* make sure that the students' learning goes in the right direction: a direction that points towards the purpose of the activity.

In a transactional perspective (see Figure 3.2), the mechanism of learning is described as the creation of relations between what *stands fast* for the student –

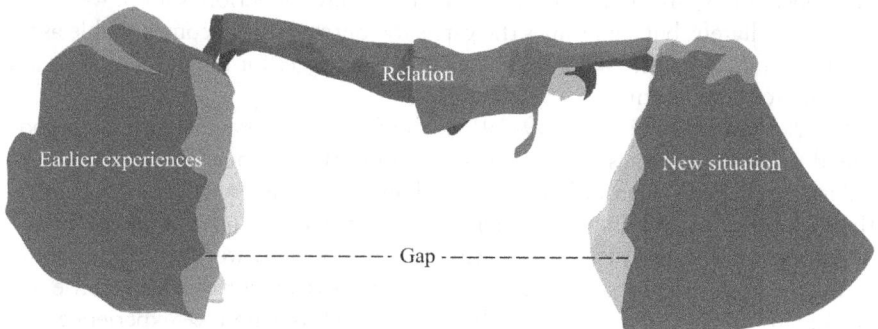

Figure 3.2 The figure illustrates the transactional perspective on learning: bridging of a gap by creating a relation between earlier experience and the new situation encountered

Source: Author's own.

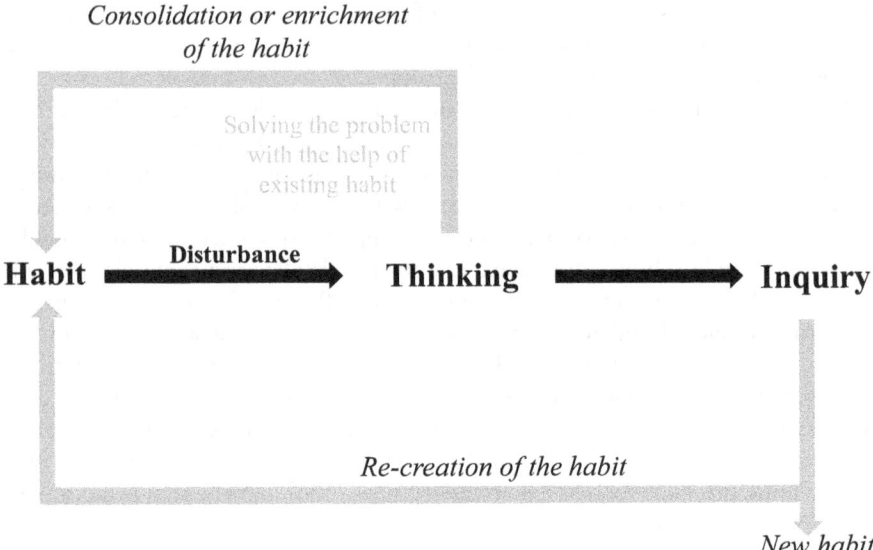

Figure 3.3 The process of consolidating, transforming and creating a new habit
Source: Author's own.

students' old experience (earlier acquired knowledge, skills, values, etc.) – and *the new situation* that the student is encountering (Wickman and Östman 2002). If the student manages to create a *relation*, it means that s/he has been able to use her/his experience to make the new situation intelligible. If that does not happen, the student cannot continue and might ask for help. Every time we encounter a new situation, a *gap* is occurring for us. If we manage to bridge that gap by creating a relation to what stands fast for us, we have learned something, i.e. we developed an expanded and more specific repertoire for action. Often, we bridge gaps immediately, but sometimes the gap is too big and this becomes visible as we hesitate, start to guess, ask for help, etc. This is a trigger for inquiry and, thus, for a more profound learning.

Charles Sanders Peirce (1839–1914), the founder of pragmatism and a good friend of William James, suggested that we mostly live our lives according to habituated beliefs (Ryan 2011). As *habits* dominate our everyday preoccupations, the default mode of human experience is non-reflective rather than reflective. Thinking is something that we mainly start to do when we need to get out of troubles, i.e. in situations when habits no longer give an automatic response to a change in the environment or a wish, need, etc. Thus, when we experience gaps that are too big, we start to think actively. The well-known expression from the famous Wittgenstein illustrates how essential habits are in our everyday life:

> Why do I not satisfy myself that I have two feet when I want to get up from a chair? There is no way. I simply don't. This is how I act.
>
> (1969, §148)

Once I have learned that I have two feet, I do not pay attention to them in my undertakings: I take them for granted and in such a case we bridge that gap immediately, without thinking. Once I have learned that it is cruel to harm animals, I live my life accordingly without reflecting upon it when I encounter animals. But sometimes we can come into a situation where our habits become disturbed – we can no longer bridge the gap immediately – and we stop the activity we are engaged in and instead start to reflect, for example on the question of whether it is only the animal or also its living environment that we should respect and take care of. If we are successful in our reflection, we will be able to bridge the gap. Going back to our initial example, because of education and the norms in use in the school, it can become a habit for the students to use only glass bottles. When parents question the habit, it can serve as disturbance of that habit which makes the students reflect – and perhaps change the habit.

Some of our habits can be characterised as frozen and others as plastic. *Frozen habits* can be called routines. Routines are rigid and inflexible. They result from 'repeated responses to recurrent stimuli that . . . fix a habit of acting in a certain way' (Dewey 1916, 29). As we become unaware of them and do not control them, they do not allow us to respond to changing surroundings. Routine does not prepare us for surprises or the possibilities for learning and thereby the development of the habit is limited in scope.

When it comes to *plastic habits*, the possibility for adjusting our actions in relation to changing surroundings and aims are bigger and so are the possibilities for learning. Such habits are characterised by meaningful thought, observation and reflection, which are prerequisites for a varied and elastic use in accordance with self-selected desires and purposes. They bring about the power to control the environment and to modify actions based on the results of prior experiences.

For Dewey, an inquiry starts when a break in our way of acting habitually is happening: the break wakes us up from non-reflectivity and a problematic situation occurs. We cannot immediately act, and this becomes visible in the way we hesitate, sigh, etc. A crucial step in an inquiry is to define the problem and through experimentation try to solve the problem. If we are successful, the result is new knowledge, skills, values and emotions. These new reflective outcomes consolidate or re-create a habit or become the start of a new habit.

Summary and conclusion

In this chapter, we have elaborated how the specificity of sustainability issues offers specific challenges for teaching and learning. We highlighted the need to move beyond traditional 'schooling' practices and formulated five principles for designing sustainable development teaching. These principles as well as the frameworks and models presented in the remainder of this book are based on a pragmatist, transactional didactic theory that understands learning and teaching in terms of action. Here learning is approached as a process of meaning-making that takes into account both prior experiences and the specificity of a particular learning situation. Learning results in a more developed and specific repertoire for coordinating activities with the surrounding world. It is about extended possibilities to act. A vital aspect of learning is inquiry, which involves both action and reflection as inseparable activities.

Understood this way, education can be seen as holding great potential to empower people by developing action competence as democratic subjects who actively take part in democratic processes on sustainability issues. Hence, this perspective on teaching and learning offers a fruitful framework for addressing the earlier elaborated challenges for education about sustainability issues. And, not least important, it makes it possible to stress pluralism and the ethical and political dimension of sustainability issues without falling into an anything-goes attitude.

Acknowledgements

The research on which this chapter is based on was supported by Swedish Research Council (Manners of teaching about controversial sustainability issues and students learning – Grant 2017–03662).

References

Boeve-de Pauw, J. and Van Petegem, P. (2018) "Eco-school evaluation beyond labels: The impact of environmental policy, didactics and nature at school on student outcomes". *Environmental Education Research*, 24(9): 1250–1267.

Chambers, J.M. (2008) "Human/nature discourse in environmental science education resources". *Canadian Journal of Environmental Education*, 13(1): 107–121.

Dewey, J. (1916/1997) *Democracy and Education: An Introduction into the Philosophy of Education*. The Free Press, New York.

Dewey, J. (1934/2005) *Art as Experience*. Perigee, New York.

Dewey, J. (1938/1997) *Experience and Education*. Touchstone, New York.

Dewey, J. and Bentley, A.F. (1949/1991) *Knowing and the Known*. Southern Illinois University Press, Carbondale.

Garrison, J. (2001) "An introduction to Dewey's theory of functional 'trans-action': An alternative paradigm for activity theory". *Mind, Culture, and Activity*, 8(4): 275–296.

Jakobson, B. and Wickman, P.-O. (2008) "The roles of aesthetic experience in elementary school science". *Research in Science Education*, 38: 45–65.

James, W. (1905/1961) *The Varieties of Religious Experience*. Collier Books, New York.

Kollmuss, A. and Agyeman, J. (2002) "Mind the gap: Why do people act environmentally and what are the barriers to pro-environmental behavior?" *Environmental Education Research*, 8(3): 239–260.

Ojala, M. (2012) "How do children cope with global climate change? Coping strategies, engagement, and well-being". *Journal of Environmental Psychology*, 32: 225–233.

Ojala, M. (2016) "Facing anxiety in climate change education: From therapeutic practice to hopeful transgressive learning". *Canadian Journal of Environmental Education*, 21: 41–56.

Östman, L. (1995) *Socialisation och mening: No-utbildning som politiskt och miljömoraliskt problem*. [Meaning and socialisation: Science education as a political and environmental-ethical problem]. Uppsala Studies in Education 61, Uppsala.

Östman, L. (1996) "Discourses, discursive meanings and socialisation in chemistry education". *Journal of Curriculum Studies*, 28(1): 37–55.

Östman, L. (1998) "How companion meanings are expressed by science education discourse". In Roberts, D.A. and Östman, L. eds., *Problems of Meaning in Science Curriculum*. Teachers College Press, New York, 54–70.

Östman, L. (2010) "Education for sustainable development and normativity: A transactional analysis of moral meaning-making and companion meanings in classroom communication". *Environmental Education Research*, 16(1): 75–93.

Östman, L. and Öhman, J. (2010) *A Transactional Approach to Learning*. Paper presented at John Dewey Society, AERA Annual Meeting in Denver, Colorado.

Östman, L., Svanberg, S. and Aaro Östman, E. (2013) *From Vision to Lesson: Education for Sustainable Development in Practise*. WWF, Stockholm.

Piaget, J. (1989) *The Child's Conception of the World*. Rowman & Littlefield, Totowa.

Rorty, R. (1982) *Consequences of Pragmatism: Essays 1972–1980*. University of Minnesota Press, Minneapolis.

Ryan, F.X. (2011) *Seeing Together: Mind, Matter, and the Experimental Outlook of John Dewey and Arthur F Bentley*. The American Institute for Economic Research, Great Barrington, MA.

Van Poeck, K. (2018) "Environmental and sustainability education in a post-truth era. An exploration of epistemology and didactics beyond the objectivism-relativism dualism". *Environmental Education Research*, pre-published online: https://doi.org/10.1080/13504 622.2018.1496404

Vygotsky, L.S. (1978) *Mind in Society*. Harvard University Press, Cambridge, MA.

Wickman, P.-O. (2006) *Aesthetic Experience in Science Education: Learning and Meaning-Making as Situated Talk and Action*. Lawrence Erlbaum, London.

Wickman, P.-O. and Östman, L. (2002) "Learning as discourse change: A sociocultural mechanism". *Science Education*, 86(5): 601–623.

Wittgenstein, L. (1969/1997) *On Certainty*. Blackwell, Oxford.

Wittgenstein, L. (1993) "A lecture on ethics". In Klagge, J.C. and Nordmann, A. eds., *Philosophical Occasions 1912–1951*. Hackett, Indianapolis and Cambridge, 37–44.

Part II

Choosing teaching content and approaches

4 Sustainable development teaching in view of qualification, socialisation and person-formation

Katrien Van Poeck and Leif Östman

Introduction

In both policy documents and research literature, one can read that teaching is supposed to achieve three functions, three overall purposes: qualification, socialisation and person-formation. A big challenge that we face as teachers is how to design and organise our teaching in order to include all three functions. A common way to do so is to divide these functions and to designate a specific place and time for handling each of them separately in the teaching practice. Here, we present another model for how to organise the teaching – an integrative model – which is built upon findings from learning research. Illustrating this model with concrete examples, we show how each of the three functions is vital for sustainable development teaching and how they are always interrelated in teaching activities. They function as 'companions'. Every teaching activity has a certain function in the forefront – for instance, qualification through the transfer of knowledge – but other functions are always at play in the background. That can be, for example, socialisation through implicitly offering particular values related to views of nature underlying the knowledge content. Such 'companion meanings' are very important to be aware of while developing lessons and courses and selecting subject matter. After all, it is the specific interplay between foregrounded purposes and content on the one hand and implicit companion meanings on the other that constitutes environmental and sustainability education (ESE) practice.

Functions of education

In this chapter, we will illustrate how we can plan teaching activities in order to handle qualification, socialisation and person-formation simultaneously in an ESE activity. The presented model is inspired by educational research (Bourdieu and Passeron 1970; Apple 1979; Englund 1986; Popkewitz 1992; Östman 1995; Östman 1996; Simons and Masschelein 2010) on the distinction between different 'functions of education' (Biesta 2009) and how this has been applied in ESE literature (e.g. Hasslöf and Malmberg 2014; Andersson 2018; Håkansson and Östman 2018). We start by explaining and illustrating the different functions of education and then return to the integrative model and the notion of companion meanings.

Qualification

An important function of education is qualification (Englund 1986; Biesta 2009). Education, then, is a matter of preparation: It is about equipping students with knowledge, skills and understandings that prepare them for a (future) role or task (Håkansson and Östman 2018). Qualification can take many different forms. It can be very specific – e.g. preparation for a particular job or profession, or training for specific skills or techniques such as driving a car – as well as much more general – e.g. teaching life skills, such as making sure that everybody can read and write, or other basic skills for participating in society. Most teaching has an explicit focus on qualification. It is one of the major functions and an important rationale for organised education which is often, but not exclusively, connected to economic arguments. Education thus plays a role in preparing the workforce and, hence, contributing to economic development. Beyond the world of work, however, qualifying students with knowledge and skills is also important for other aspects of our lives. It can also entail, for instance, enhancing students' cultural and political literacy by fostering knowledge and skills needed for their functioning in society or for playing their role as a citizen.

Also in an ESE context, a focus on qualification is omnipresent and takes many different shapes. An example closely connected to the economic argument is the United Nations' Economic Commission for Europe's (UNECE 2011) emphasis on the important role of education to contribute to a shift towards a green economy by equipping 'people with the values, competences, knowledge and skills that are necessary for them to put the green economy concept into practice', for instance 'skills for green jobs'. Another example of a focus on qualification, beyond the world of work, is the argument underpinning a 'fact-based' approach to ESE (see chapter 5) that students need to acquire basic knowledge in the natural sciences because it can prepare them for understanding the limits of earth. Or the design of educational practices aimed at teaching people which kind of behaviour can contribute to reducing their ecological footprints. Framed in terms of 'learning objectives', a focus on qualification is illustrated by the following examples related to different Sustainable Development Goals (SDGs) (UNESCO 2017):

- 'The learner understands how extremes of poverty and extremes of wealth affect basic human rights and needs.'

  (p. 12)

- 'The learner is able to communicate on the issues and connections between combating hunger and promoting sustainable agriculture and improved nutrition.'

  (p. 14)

- 'The learner is able to include health promoting behaviours in their daily routines.'

(p. 16)

- 'The learner understands the concept of gender, gender equality and gender discrimination and knows about all forms of gender discrimination, violence and inequality and understands the current and historical causes of gender inequality.'

(p. 20)

- 'The learner is able to plan, implement, evaluate and replicate activities that contribute to increasing water quality and safety.'

(p. 22)

All these objectives aim to equip students with knowledge, skills and understandings that prepare them for contributing to realising SDGs such as overcoming poverty, ending hunger and promoting sustainable agriculture, ensuring good health and well-being, achieving gender equality and ensuring clean water and sanitation.

Socialisation

Another frequently described function of education is socialisation: teaching students certain values, attitudes, norms and worldviews so as to socialise them into the prevailing standards of a particular social, cultural and political order or group (Durkheim 1911; Bourdieu and Passeron 1970; Apple 1979; Östman 1996; Biesta 2009; Håkansson and Östman 2018). Socialisation is, without doubt, one of the actual 'effects' of education. After all, teaching and learning are never neutral and always involve the representation of norms and values (see chapter 15). Sometimes socialisation is foregrounded and actively pursued by teachers, educational institutions or educational policies. The goal, then, is the transmission of particular norms and values, for instance for the purpose of the continuation of cultural or religious traditions or for the introduction of students into a professional community or their initiation into democratic values and practices. But even if socialisation is not the explicit aim, educational practices will often – if not always – have a socialising effect (see the section on 'Companion meanings').

In relation to ESE, a focus on socialisation is reflected in aims and attempts to transmit values, attitudes, norms and worldviews that are considered vital for sustainable development, e.g. human dignity and rights, social equality, gender equity, care for the environment, solidarity, cultural diversity, participation, economic vitality, justice, social cohesion and mutual respect between people, countries and generations (UNECE 2005; UNESCO 2005). The pursuit of a more sustainable world, it is argued, requires that 'everyone . . . learn[s] the values, behaviour and lifestyles required for a sustainable future and for positive societal transformation' (UNESCO 2005, 6). For instance, when UNECE (2011) argues for the importance of 'a shift in economic thinking' by changing people's mindset, the role of education in the context of a green economy is clearly seen in terms of socialisation – besides its focus on qualification, described earlier. A prerequisite for achieving a shift in economic thinking, it is argued, is 'a reorientation of values and attitudes' of a broad variety of actors (individual consumers, policymakers, multinational corporations, etc.). Education is attributed a unique and crucial role in making this possible if it is 'designed in a way to facilitate the development of values and to initiate the reconsideration of existing values and attitudes'.

Several of the learning objectives that UNESCO (2017) considers key for realising the SDGs have socialisation as a purpose, for instance,

- 'The learner is able to feel empathy and solidarity with those who differ from personal or community gender expectations and roles.'

(p. 20)

- 'The learner is able to feel empathy for and to show solidarity with people who are discriminated against.'

(p. 30)

- 'The learner is able to maintain a vision of a just and equal world.'

(p. 30)

- 'The learner is able to feel responsible for the environmental and social impacts of their own individual lifestyle.'

(p. 32)

Socialisation, here, takes the form of transmitting (a) values such as equality, solidarity and justice, (b) attitudes such as feeling empathy and responsibility and (c) norms and worldviews related to a just and equal world.

Person-formation

A third important function of education, person-formation (Popkewitz 1992; Håkansson and Östman 2018), is connected to the formation of the self: the cultivation of people's personalities or, in other words, the process of personal maturation. This function of education is closely related to what is often called 'Bildung' in educational literature: a form of general education aimed at personal growth and the development of individual agency and qualities such as critical thinking or moral judgement. Bildung is distinguished from mere training in specific skills, e.g. vocational training (*Ausbildung*). Important to realise is that person-formation is seen as a life-long process that takes place through everyday actions that shape people's ways of acting, thinking and being (see chapter 15). Obviously, this also happens in the classroom. Students 'become someone' in relation to specific teaching practices in which certain knowledge is made available and through which a certain spectrum of possible action is created.

Sometimes, person-formation is an explicit goal of a lesson, course or curriculum. This is illustrated by some of the 'cross-cutting key competencies' for education in relation to the SDGs formulated by UNESCO (2017, 10) that highlight personal growth through moral judgement and critical thinking as an important aspect of ESE:

- 'Normative competency: the abilities to understand and reflect on the norms and values that underlie one's actions, and to negotiate sustainability values, principles, goals, and targets, in a context of conflicts of interests and trade-offs, uncertain knowledge and contradictions.'
- 'Critical thinking competency: the ability to question norms, practices and opinions, to reflect on own one's values, perceptions and actions, and to take a position in the sustainability discourse.'

Translated into more concrete learning objectives, this becomes, for instance:

- 'The learner is able to reflect on their own gender identity and gender roles.'
  (p. 20)

- 'The learner is able to reflect on their own personal belonging to diverse groups (gender, social, economic, political, ethnical, national, ability, sexual orientation etc.) their access to justice and their shared sense of humanity.'
  (p. 42)

In teaching and learning practices, the amount of room for students to think and act for themselves can differ largely. As a result, person-formation can take different forms. Often, it takes the form of *identification*, i.e. the development of a specific identity, which shows some similarities and overlap with socialisation. Identification often occurs in relation to the content of qualification and socialisation. When, for instance, learning dos and don'ts in view of reducing one's ecological footprint, students can simultaneously develop an identity, e.g. of an ecological citizen, through an identification process within the presented discourses, practices and habits. Ideland and Malmberg (2015, 173), for instance, have analysed teaching materials used in ESE and showed how a hypothetical 'desirable child', i.e. an 'eco-certified child', is constructed in these materials. In the textbooks, games and children's books they have analysed, certain characteristics and activities stand out as normal and desirable, thus governing in a very specific way what it means to be environmentally friendly. Many of the presented discourses, practices and habits have in common that they are framed in a logic of scientific and mathematical objectivity and faith in technological development and in market-driven solutions. Hence, through these teaching and learning activities, students are encouraged to develop specific, 'eco-certified' identities, for example being a 'well-informed and curious consumer'. This shows how every way of teaching a particular content simultaneously offers students certain 'subject positions' that allow them to identify with a specific role (see chapter 15). This is also the case during activities that, at first sight, seem to be very 'neutral' such as studying a biotope (Van Poeck et al. 2016).

During an excursion of secondary school students to a nature education centre, the educator's purpose in the forefront was to teach the students about water quality and, in particular, how to assess this by determining the biotic index.[1] The students were given the task to find and identify aquatic animals using a dip net, magnifying pots and identification tables in order to measure the biotic index. The educator frequently intervened with instructions for the students to select the 'proper' animals, i.e. macroinvertebrates needed to calculate the value of the biotic index and not, for instance, little fish and tadpoles that made several students enthusiastic and excited when they collected them in the net. Thus, along with fact-based knowledge on biological water quality, the students were also offered a specific identity. The message they were offered in the background is that studying biology requires that they become skilled scientific 'manipulators', i.e. that they experiment (catch, measure, etc.) with animals in order to create scientific knowledge (see also Östman 2007).

Besides such identification, person-formation can also take the form of *subjectification* (Biesta 2009; Simons and Masschelein 2010; Van Poeck and Vandenabeele 2012). This is a process of individuation, of developing mature and independent ways of being and acting in the world, as a 'subject' instead of the 'object' of other people's purposes and desires. Thus, in contrast to identification, subjectification can best be understood as the *opposite* of socialisation. Whereas identification is about embracing a specific, pre-existing role or subject position offered through the content of qualification and socialisation, subjectification is about developing into a unique person beyond such offered roles. It involves 'dis-identification' (Rancière 1995, 1999): it disrupts the discourses, practices and habits in which one is participating by establishing a subjectivity, by embracing a way of being that has no place in the existing order of things and hints at independence from it (Håkansson and Östman 2018). Through subjectification, new ways of doing and being come into existence and the taken for granted is disrupted. Person-formation is thus a complex process of identification and dis-identification through an ambiguous interplay between taking up pre-existing identities in line with particular subject positions and developing one's singular subjectivity. In this interplay, identification with practices and roles that students encounter at school can involve a dis-identification from the ones that they are familiar with at home.

Subjectification can take place anytime, often unexpectedly, in response to encounters with existing practices and roles. However, as teachers, we can also deliberately create opportunities for it by offering students room to think and act for themselves, for example through values-clarification exercises (see chapters 18 and 19) that encourage them to investigate and reveal different perspectives on an issue, to think critically about alternative options and sometimes also to change their own perspective (Hasslöf and Malmberg 2015). As described in the step-by-step guidelines for the values-clarification exercise 'Four corners' (see chapter 19 – Textbox 19.1), the opportunity for students to change their position is a vital aspect of such activities. This can open up possibilities to dis-identify from well-established practices and roles within the school, family, etc. by changing one's perspective.[2]

Besides such a shift of perspective, subjectification can also take place when people are dislocated by an experience that disrupts the ground on which their existing practices, roles and habits are built.[3] Such instances of subjectification move beyond the cognitive dimension of factual negotiations or being convinced by others' rational arguments. Instead, it largely involves personally felt emotions, values, authentic beliefs, etc. and can thus be seen as a very existential experience. It mostly occurs unexpectedly and is difficult, if not impossible, to plan or guarantee. The following example (see also chapter 10) shows how it can take place in response to an activity where university students are socialised into the role of natural scientists and are thereby offered a specific role to identify with, i.e. a rational manipulator of nature (see earlier). A group of biology students are given the task to collect animals on the shore and to put them in an aquarium for further scientific studies. While picking up a sea urchin and feeling the resistance as the animal attaches to a stone with suction cups, one of the students reacts as follows, 'Hell, for crying out loud. It feels awful when you pull them loose . . . . It seems weird. We've got to learn to pick them off with our hands'. This spontaneous expression of dislike reveals a strongly felt emotion in the midst of preparation for scientific study. The student tells the teacher that she does not want to continue gathering animals. She discovered a new self and does not want to take on the role of a manipulator of nature. When this kind of subjectification emerges in education practice, it offers great opportunities for teachers to develop, with the whole class, experience-based and creative discussions about existential, moral and political aspects of our relation to our environment. This can be a bit risky in classroom practices considering the strong emotions it may evoke (see chapters 8 and 20), but it is also an important driver for change and transformation – a prerequisite for teaching in view of a more sustainable world (Lotz-Sisitka et al. 2015). As Garrison et al. (2015, 196) argue, not only critical thinking but also creativity are vital aspects of ESE: '[often] critical thinking confines itself to simply choosing among pre-existing alternatives instead of imagining or creating new desirable . . . possibilities for the future'.

The distinction between qualification, socialisation and person-formation as summarised in Table 4.1 serves as a framework for the integrative model for organising teaching where all functions are taken into account simultaneously. Important to realise – as the above presented examples show – is that teaching and learning knowledge is always accompanied by teaching and learning values and that the knowledge content of textbooks and lessons always also functions implicitly as socialisation and identification/subjectification content. The implicit holds strong potential to (re-)create values, worldviews and identities and to affect how students perceive themselves in relation to nature, to their fellow human beings, etc. Thus, it contributes to justifying, strengthening or, otherwise, questioning the status quo in society. We will further elaborate on this in the next and final section.

Table 4.1 Functions of education

| | Qualification | Socialisation | Person-formation | |
			Identification	Subjectification
Purpose of education	Equipping students with skills and competences that prepare them for a role of task	Transmitting values, attitudes, norms and worldviews	Taking on an identity in relation to the content of preparation and socialisation	Becoming a subject, developing new ways of doing and being
Examples	Learning to reduce one's ecological footprint, learning skills for jobs in the green economy, learning knowledge about ecosystems . . .	Learning sustainable values and attitudes, learning to feel responsible for the ecological and social impact of one's lifestyle . . .	Becoming an environmentalist, a sustainable citizen, a business person, a well-informed consumer . . .	Change one's perspective on what it means to be a business person, taking an emotionally invested political stand . . .

Companion meanings and the integrative model

As indicated, qualification, socialisation and person-formation are companions. They come together in educational practice. Inspired by Dewey's (1938) notion of collateral learning, Roberts and Östman (1998) introduced the term 'companion meaning' for understanding and investigating how every teaching activity has a certain meaning-making in the forefront while there are always other meanings that follow automatically, in the background: companion meanings. Empirical investigations of science textbooks, for instance, reveal how the presented knowledge about the world is always entangled with specific views of nature and how to treat it (Östman 1996). A lesson about atomic theory, for instance, can have the transfer of scientific knowledge of atoms, molecules, etc. as its main goal. Yet, a particular understanding of nature automatically accompanies these scientific meanings, that is, the atomistic and deterministic view of nature that has been developed since 1600 and largely affects our society up to today (Worster 1985). As such, the qualification goal of developing particular scientific competences simultaneously goes with socialisation. Students are simultaneously introduced into the specific values, attitudes, norms and worldview characterising modern society. At the same time, they will automatically form – or re-form – their identity in relation to this learning content, for instance by taking on the identity of 'an objective, rational and neutral scientist'. Another example is that when a chemistry textbook contains knowledge on how air can be used to produce raw materials for utilisation in various operations, nature is implicitly represented as something that can and should be exploited for promoting human beings' material welfare. Thus, teaching that places qualification in the forefront also implies socialisation and identification in the background, by offering certain companion meanings. Similar mechanisms are at play in the examples discussed in this chapter. When a teacher aims to qualify students 'to plan, implement, evaluate and replicate activities that contribute to increasing water quality and safety' (UNESCO 2017, 22), for instance, nature emerges as something that can and should be taken care of by human beings. Or by learning skills for the green economy, students are offered a particular worldview in line with the discourse of 'ecological modernisation', i.e. the dominant way of conceptualising environmental problems in the Western world since the late 1970s that is built on the idea that it is possible to reconcile economic growth, technological development and the solution of ecological problems (Hajer 1995). Or via games, web-based calculators and textbooks on the ecological footprint (Ideland and Malmberg 2015; Öhman et al. 2016; Andrée et al. 2018), students are not merely offered knowledge on how to reduce their ecological impact but are, simultaneously, offered particular epistemological values – i.e. that the question of what is ecologically desirable has an objective answer based on the proper calculations – as well as specific identification content – i.e. that becoming a responsible person is achieved through individual consumption choices rather than taking the issues to the political level.

These and the above examples show our teaching is always an integrated activity. When we teach knowledge, we simultaneously offer students very particular worldviews – ways of reasoning about the world and their place in it – as well as specific roles. Every teaching activity inevitably involves choices about what to put in the forefront and what to background. It also requires making judgements about whether the companion meanings offered in the background are in accordance with the explicit teaching goals. With the aid of the integrative model, we can as teachers consciously take control of the companion meanings that we are offering students and thereby deliberately plan for all three functions within one and the same lesson or series of lessons.

Acknowledgements

The research on which this chapter is based on was supported by Ghent University (International Thematic Network 'SEDwise – Sustainability Education: Teaching and learning in the face of wicked socio-ecological problems') and by Research Foundation Flanders (Scientific Research Network 'Public pedagogy and sustainability challenges').

Notes

1 The biotic index is a scale for the quality of an environment – often used to assess water quality – by indicating the types of organisms present in it. It works by assigning different

levels of tolerance to pollution to the different types of organisms. Based on the identified organisms, the water quality can be determined as excellent, good, fair or poor.

2 Elsewhere, we have called this form of subjectification 'subjectification as perspective-shifting' (Håkansson and Östman 2018).

3 This form of subjectification we have called 'subjectification as dismantling' (Håkansson and Östman 2018).

References

Andersson, P. (2018) "Business as un-usual through dislocatory moments: Change for sustainability and scope for subjectivity in classroom practice". *Environmental Education Research*, 24(5): 648–662.

Andrée, M., Hansson, L. and Ideland, M. (2018) "Political rationalities in science education: A case study of teaching materials provided by external actors". In Otrell-Cass, K., Orlander, A.A. and Sillasen, M.K. eds., *Cultural, Social, and Political Perspectives in Science Education*. Springer, Cham, 75–92.

Apple, M. (1979) *Ideology and Curriculum*. Routledge & Kegan Paul, London.

Biesta, G. (2009) "Good education in an age of measurement: On the need to reconnect with the question of purpose in education". *Educational Assessment, Evaluation and Accountability*, 21(1): 33–46.

Bourdieu, P. and Passeron, J.C. (1970) *La reproduction. Éléments pour une théorie du système d'enseignement*. Editions de Minuit, Paris.

Dewey, J. (1938/1997) *Experience and Education*. Touchstone, New York.

Durkheim, E. (1981/1911) "De educatie, haar aard en haar rol". In Du Bois-Reymond, M. and Wesselingh, A. eds., *School en maatschappij. Sociologen over onderwijs en opvoeding*. Wolters-Noordhoff, Groningen, 209–223.

Englund, T. (1986) *Curriculum as a Political Problem. Changing Educational Conceptions, with Special Reference to Citizenship Education*. Studentlitteratur/Chartwell-Bratt, Lund.

Garrison, J., Östman, L. and Håkansson, M. (2015) "The creative use of companion values in environmental education and education for sustainable development: Exploring the educative moment". *Environmental Education Research*, 21(2): 183–204.

Hajer, M. (1995) *The Politics of Environmental Discourse: Ecological Modernization and the Policy Process*. Oxford University Press, New York.

Hasslöf, H. and Malmberg. C. (2014) "Critical thinking as room for subjectification in education for sustainable development". *Environmental Education Research*, 21(2): 239–255.

Håkansson, M. and Östman, L. (2018) "The political dimension in ESE: The construction of a political moment model for analyzing bodily anchored political emotions in teaching and learning of the political dimension". *Environmental Education Research*, pre-published online: https://doi.org/10.1080/13504622.2017.1422113

Hasslöf, H. and Malmberg, C. (2014) "Critical thinking as room for subjectification in education for sustainable development". *Environmental Education Research*, 21(2): 239–255.

Ideland, M. and Malmberg, C. (2015) "Governing 'eco-certified children' through pastoral power: Critical perspectives on education for sustainable development". *Environmental Education Research*, 21(2): 173–182.

Lotz-Sisitka, H., Wals, A., Kronlid, D. and McGarry, D. (2015) "Transformative, transgressive social learning: Rethinking higher education pedagogy in times of systemic global dysfunction". *Current Opinion in Environmental Sustainability*, 16: 73–80.

Öhman, J., Öhman, M. and Sandell, K. (2016) "Outdoor recreation in exergames: A new step in the detachment from nature?" *Journal of Adventure Education and Outdoor Learning*, 16(4): 285–302.

Östman, L. (1995) *Socialisation och mening: No-utbildning som politiskt och miljömoraliskt problem* [Meaning and socialisation: Science education as a political and environmental-ethical problem]. Uppsala Studies in Education 61, Uppsala.

Östman, L. (1996) "Discourses, discursive meanings and socialisation in chemistry education". *Journal of Curriculum Studies*, 28(1): 37–55.

Östman, L. (2007) "Content analysis, curriculum theory and didactics: Science education as a political and environmental-ethical problem". In Forsberg, E. ed., *Curriculum Theory Revisited*. Uppsala University, Studies in Educational Policy and Educational Philosophy, Research Reports 10.

Popkewitz, T.S. (1992) *The Study of History, Social Regulation and Power in Schooling*. Presentation at the Educational Department, Uppsala University.

Rancière, J. (1995) *On the Shores of Politics*. Verso, London and New York.

Rancière, J. (1999) *Dis-agreement: Politics and Philosophy*. University of Minnesota Press, Minneapolis and London.

Roberts, D. and Östman, L. eds. (1998) *Problems of Meaning in Science Curriculum*. Teachers College Press, London.

Simons, M. and J. Masschelein. 2010. "Governmental, political and pedagogic subjectivation: Foucault with Rancière". *Educational Philosophy and Theory*, 42(5/6): 588–605.

United Nations Economic Commission for Europe – UNECE. (2005) *UNECE Strategy for Education for Sustainable Development Adopted at the High-Level Meeting of Environment and Education Ministries*. (Vilnius, 17–18 March 2005) UN, New York.

United Nations Economic Commission for Europe – UNECE. (2011) *Discussion Paper on the Role of Education for Sustainable Development in Shifting to a Green Economy*. UN, New York.

United Nations Educational, Scientific and Cultural Organization – UNESCO. (2005) *United Nations Decade of Education for Sustainable Development (2005–2014): International Implementation Scheme*. UNESCO, Paris.

United Nations Educational, Scientific and Cultural Organisation – UNESCO. (2017) *Education for Sustainable Development Goals. Learning Objectives*. UNESCO, Paris.

Van Poeck, K., Goeminne, G. and Vandenabeele, J. (2016) "Revisiting the democratic paradox of environmental and sustainability education: Sustainability issues as matters of concern". *Environmental Education Research*, 22(6): 806–826.

Van Poeck, K. and Vandenabeele, J. (2012) "Learning from sustainable development: Education in the light of public issues". *Environmental Education Research*, 18(4): 541–552.

Worster, D. (1985) *Nature's Economy: A History of Ecological Ideas*. Cambridge University Press, Cambridge.

5 Different teaching traditions in environmental and sustainability education

Johan Öhman and Leif Östman

Introduction

Research into the history of school subjects has shown that different traditions of how material and methods are selected in education are present within different school subjects. These traditions can therefore be termed 'selective traditions'. The term was originally developed by Williams (1973) to underline the fact that a certain approach towards knowledge and a certain educational praxis is always selected within the framework of a specific culture. The regular patterns of the selective processes that develop over time form a selective tradition. These selective traditions represent a number of solutions as to what constitutes the best form of teaching within a subject, and they also include different approaches in the choice and organisation of content and the choice of teaching methods. Like most established traditions, selective traditions often function as an oblivious frame of reference. This may result in new educational guidelines and objectives being interpreted within the frames of prevalent traditions. Here there is a great risk that new ideas will never be realised. It is therefore important for teachers to be aware of the traditions that exist within a subject or field of study in order to be able to make critical and conscious choices of educational content and methods.

In this chapter, we will describe three traditions prevailing in environmental and sustainability education. The purpose of clarifying these traditions here is to establish a reference point that can be applied when discussing teaching that involves issues relating to the environment and sustainable development. They can be seen as alternatives to reflect on, oppose or support when planning lessons or formulating ideas. We will also discuss the strengths and shortcomings of the traditions in relation to two interconnected premises: that environmental and sustainability issues are *value issues* and that they should be dealt with *democratically*.

Selective traditions in environmental and sustainability education

Previous research has shown that there are reasons to talk about three different selective traditions within ESE (Öhman 2004; Sandell et al. 2005; Öhman 2008; Rudsberg and Öhman 2010; Östman et al. 2013). Here these teaching traditions are labelled according to their main orientations: *fact-based*, *normative* and *pluralistic*. This categorisation of ESE into three separate traditions is originally based on the results from the Swedish National Agency for Education's evaluation of environmental education in schools (from pre-schools to upper secondary schools in all subjects) in 2001 led by Leif Östman and Johan Öhman (Swedish National Agency for Education 2001). The analyses of data coming from a national survey and teacher interviews identified the division of three separate traditions and demonstrated that they were logical and coherent. The traditions differ in four fundamental approaches: their *sustainability approach*, *didactic approach*, *approach to facts and values* and *approach to democracy and education*.

The *sustainability approach* concerns how the causes, character, extent and seriousness of environmental and sustainability problems are perceived and what the main solutions to these problems are claimed to be (see chapter 1). Taking the conflicting interests that can appear in environmental and sustainability issues into account, it is easy to understand why people adopt opposing positions in the wide range of topics related to sustainable development based on their perspectives, values and ideologies. For example, this concerns questions about how much trust to put in scientific and technological solutions, the extent to which changes in individual lives are necessary and whether it is believed that solutions exist within the current economic and social system (the liberal market economy) or if a radical change of system is necessary. An essential aspect is which of the ecological, economic and social dimensions are considered to be the most important and the pivotal aspect for creating sustainable development.

The traditions also differ in the ways educational practice in general is perceived and conducted; we refer to this as *didactic approach*. Didactic approach encompasses general ideas about the role and purpose of schools in society, as well as ideas which directly affect the teaching process. One way of structuring an understanding of didactic approach is to start from the three main questions in education of *why?* – the motives of education, *what?* – the content of education and *how?* – the methods used in education. There are many different answers to each of these questions.

At a social level, the *why* question addresses the main function of the school – the purpose the school serves in society. There is a basic difference of opinion here between those who regard the school as an institution that supports continuity in society, in the sense that it preserves and cultivates basic norms, values, viewpoints and knowledge, and those who suggest that the primary function of schools is to create change in society by questioning and critically assessing that which is considered habitual and taken for granted. The *why* question therefore encompasses visions of an ideal society – not least how we see the role of the school in a democracy. Based on these broad standings on the role of the school in society, it is possible to observe differing opinions on the motives of teaching in various subjects and the kinds of competence students should develop.

The *what* question addresses the choice of content and the particular grounds on which certain material is chosen. Every section of the curriculum can potentially include a great deal, even within the limits of school policies. The question therefore is: What is most important and central in each case? This question is also concerned with the reasons why a certain type of material is selected. Here, there is a basic difference between those who claim that specific content needs to be taught in schools based on the traditions in different academic disciplines and those who stress that the teaching content should be selected in relation to the needs and interests of the students and societal continuity and changes.

The *how* question deals with the choice of teaching methods. A central question here is: How can students develop and achieve the goals that have been established in the curriculum in a functional way? It therefore includes an understanding of how students' learning processes take place and how the role of the teacher is perceived. It could be said that two opposing positions have characterised the educational debate on these issues in the last century. From one position, the student is seen as a passive receiver of information who requires external motivation. The role of the teacher is that of an expert in the subject, who has the task of transferring knowledge and facts to the students. From the other position, the student is regarded as motivated, active and knowledge seeking. The educational emphasis is on cooperation and problem solving as the most important aspects of the learning process. Students' knowledge is primarily developed through first-hand experiences of the natural world and society, with the teacher being seen as a facilitator or guide in these processes.

Approach to facts and values concerns both philosophical and educational questions about the importance of facts and values and the relation between them. That is, are facts or values the focus of educational efforts, and if so, which are seen as the main solution to sustainability problems: more knowledge or changes in the way humans value things (e.g. what is important in life, responsibility for coming generations, care for nature etc.)? Relations between facts and values concern whether facts and values are separate domains or integrated, whether there is such a thing as objective facts or if all descriptions of the world are always coloured by values, and whether it is possible to derive values from facts. This aspect also concerns the responsibility of schools and education for facts and values. Everybody agrees that schools are responsible for young people's knowledge development, but when it comes to values, things are more diverse. Are schools also responsible for the socialisation of certain values? If so, which values and to what extent? How strong and what kind of normativity is acceptable in education? Is the way we value environmental and sustainable issues a private question, or is the global environmental and social crisis so serious and urgent that teachers are obliged to encourage specific values that promote sustainable development?

The value aspect of education also relates to different *approaches to democracy and education*. Central here is what has been called the *democratic paradox of education*, that is, the double educational assignment to foster free, autonomous subjects and at the same time transfer the basic values and norms of a particular culture to future generations. Different ways of handling this paradox imply different views on the role of ESE in relation to citizenship education and the relation between education and the democratic process, e.g. whether the democratic process comes *after*, comes *before* or is situated *in* the educational activities. An important aspect of the approach to democracy and education is how democracy is manifested in the teaching process. That is, the degree to which students should participate in the constitution of knowledge and values and be involved in the planning and realisation of the lessons.

The three traditions of ESE each form a logic unity of the approaches to sustainability issues, the central didactic questions, the fact-value relation and the democratic role of education. The traditions can be seen as historical accounts of environmental and sustainability teaching and can be accredited with constituting a part of the development where one tradition is based on the other. But studies have also shown that these three traditions are all present in schools today (Andersson 2017; Sund and Wickman 2011; Borg et al. 2012, 2014).

Below we take a closer look at the characteristics of the three traditions of environmental and sustainability education, with reference to their approaches to sustainability issues, the central didactic questions, the fact-value relation and democratic approach. It is important to point out that the descriptions of these traditions are in ideal form and have been edited in order to make them clearer. As we are aware, reality is considerably more complex than our descriptions of it – there are infinite details to acknowledge in the practicalities of teaching, and people are not always logical in their actions. In practice, teaching is always made up of a combination of different perspectives, although most of us teachers usually remain within the logics of one tradition.

The fact-based tradition

In the fact-based tradition, environmental and sustainability problems are regarded as *knowledge-based problems* – we have these problems because the public is ignorant and we lack knowledge about how to design the most effective actions to mitigate the problems. Environmental and sustainability problems are therefore approached as questions of science (mainly natural science). More research and technology development and supplying more information to the public are expected to lay the basis for sustainable development.

The focus in the lessons is accordingly on conveying scientific models, facts and concepts that have been adapted for educational purposes. The position taken is that only science and scientific facts are reliable for knowledge about environmental issues. Scientific facts and models therefore have a sole importance in teaching. Based on these presumed objective facts, the students are then expected to draw independent conclusions and act on them. In order to gain an understanding of environmental and sustainability problems, the teaching process focuses on subject knowledge. The most usual method of teaching in schools is by teacher-led lessons. If necessary (or possible), practical experiments are carried out in order to illustrate particular phenomena. Field trips and other excursions also take place to a certain extent. Student participation takes place to the extent that the teacher observes what students seem to be interested in and incorporates this into future lesson plans.

In the fact-based tradition, facts are separated from values. Values are regarded as subjective and, as they belong to the private sphere of the students, cannot therefore be dealt with in a rational discussion. Consequently, morals or politics should not be part of environmental and sustainability education. As objective education is a prioritised value; it is not the task of schools to exert an influence on the ethical or political stands of the students. The guarantor for an objective education is that the facts learned in school originate from scientific knowledge. Thus, this tradition is based on the conviction that the scientific method is value-free and that science provides the necessary knowledge for solving sustainability problems. The democratic role of education is to provide objective facts as a basis for students' opinion-making. By obtaining more knowledge, students will gradually become full members of the democratic community. When their education is completed, they will then be able to exercise their rights as democratic citizens. The democratic process is therefore something that comes *after* education. In this way, the fact-based tradition avoids the democratic paradox of education.

The normative tradition

Within this tradition, environmental and sustainability problems are primarily considered to be of a *moral* character and are resolved by adopting environmentally friendly and sustainable values, norms and lifestyles. The values and norms are based on scientific knowledge, where scientific knowledge is accordingly seen as having moral implications – the facts provided by science can be used to derive the good values and right norms of sustainable development. When people act on these values and norms, the whole of society can be adjusted according to contemporary scientific knowledge about the state of the world and predictions by scientists. Experts from various fields of science should therefore be those who advise and direct people in terms of how they should approach sustainability issues and which sustainable values they should adopt. To adopt a sustainable lifestyle then becomes an individual responsibility.

The answers to value-related questions are determined by means of deliberative discussions between experts and politicians based on scientific facts about the current ecological state of the world, which are then presented in policy documents and syllabuses. The democratic process is thus something that comes *before* education. This also implies that it is possible to come up with universal solutions to environmental and developmental problems. Schools are then obliged to teach students sustainable values and norms and, in this way, attempt to change their behaviour in the desired direction. In relation to the democratic paradox of education, emphasis is put on the necessity to transfer a certain mindset to the next generation.

An essential goal of this tradition is to encourage the students to commit to environmental and sustainability issues. In the educational practice, the emphasis is on developing the ability to form and defend a standpoint that is based on scientific facts. The reference point for this tradition is the idea that a strong, almost causal, relationship exists between knowledge, values and behaviour. If students are more knowledgeable about, for example, fundamental ecological conditions, they will naturally begin to act more responsibly towards the environment. A central aspect is the development of practical skills, i.e. the ability to put into everyday practice what has been studied and discussed in theory.

The lesson content includes current and local issues, with global problems and future consequences also being addressed. In addition to environmental issues, resource distribution and population growth are included. The educational content is sometimes organised in a thematic way, in which several teachers of different subjects cooperate. Although the lessons are based on scientific facts, values and emotional aspects are also regarded as important for commitment.

To ensure that the lessons achieve the intended objectives, a great deal of attention is given to working methods and using reference points that are based on the students' experiences and attitudes. The lessons are often group-based activities with a focus on problem solving, where students look for facts and information themselves, or are practically activated in other ways. Field trips are also part of this tradition, as certain aspects of the lessons require first-hand experience. The teacher and students often carry out lesson planning together.

The pluralistic tradition

The increasing uncertainty about environmental and sustainability issues and the proliferation of differing opinions in the debate on sustainable development are central points of departure of this tradition (see chapter 2). Sustainability problems are understood as conflicts between different human interests, values and ideologies. This implies that these problems are seen as *political* issues. Different groups of people with equally different viewpoints and values have their own opinions about what constitutes a problem and also have differing views about how serious these problems are. Even if people are able to agree on the facts, they might have different ideas about the most effective or suitable way of achieving sustainable development due to their differing ideological beliefs. As science is confined to supplying facts, it is not regarded as the only source of guidance when it comes to the political and ethical aspects of sustainable development.

This conflict-based approach to ESE is characterised by a striving to highlight different perspectives, views and values when dealing with questions and problems concerning the future of our world. In contrast to the fact-based tradition, facts and values as well as emotions are subjects of a rational discussion that is open and does not aim for a preconceived version of the state of the world or an ideological standpoint. The way of finding a common understanding of the relation between facts and values, or recognising and accepting our different standpoints, is seen as being accomplished by democratic discussion, i.e. deliberation. Such discussions are an essential part of education in the pluralistic tradition, and the democratic process is accordingly situated *in* education itself. We could say that the aim of pluralistic education is to enhance students' competence to critically evaluate different perspectives of environmental and developmental issues, to take a stand and participate in debates, discussions and decisions at a private everyday level and a comprehensive societal level. When it comes to the democratic paradox of education, emphasis is accordingly put on students' free will and strengthening them as autonomous subjects.

The teaching material includes the relationship between local and global problems. The focus is on sustainable development and the related topics of economy, politics and ecology. Sustainable development is a recurring theme in all subjects, due to its integrated character. This implies that perspectives from natural and social sciences, humanities and arts feature in the lesson material.

The varied character of the problems encountered in the different aspects of the lessons indicate that the teaching methods are also varied. Discussing a wide range of viewpoints is considered an important aspect of the lessons. Students critically investigate the knowledge basis that supports the various standpoints and the values and ideologies that are connected to different interest groups. From this perspective, it is important that lessons include, for example, panel debates or role-playing activities that reflect real conflicts in society and that students are permitted to participate in the sustainability aspects of the organisation of the school. The teacher is crucial for the quality of the discussions and students' learning process. It is the teacher's responsibility to make sure that the referred facts are correct and to clarify, question, problematise and make suggestions in order to stimulate further discussion.

Valuing the traditions

As a teacher, a central question is, of course, how to value these traditions. Which tradition should guide my teaching? Every tradition has its shortcomings and strengths. In order to make such a judgement, we must first decide the premises or criteria for an assessment. Based on what has been stated previously in this book (chapters 2 and 3), we will here discuss the strengths and shortcomings of the traditions in relation to two interconnected premises. The first is that environmental and sustainability issues are *value issues* that are connected to facts, interests, preferences, beliefs and attitudes. The second is that environmental and sustainability issues should be dealt with *democratically*, in that moves towards sustainable development imply considerable changes to our society that will affect every citizen's everyday life. It is therefore a major concern of schools to educate about, for and in this democratic process in order to support students' growth as democratic citizens. This process is closely related to the distribution of authority in the classroom. Different ways of distributing authority have different companion meanings (see chapter 4), which influence students' views of democracy and their role in it. Authority here refers to those who have the right to speak during a lesson and those who are simply regarded as listeners. If the intention is that students are to take active roles in social change, in their capacity as citizens, it is reasonable that they should be encouraged to learn about the various roles a citizen might have in such a debate.

Fact-based – reliable but omits values

In the fact-based tradition, sustainability problems are seen as being caused by a lack of knowledge and information. The strength of the fact-based tradition is that it is based on reliable and established scientific knowledge. A lot of time is spent learning this knowledge, which gives the students a solid basis for their decisions later in life. They also learn to understand the importance of science and the careful verification processes of scientific facts. In this way, they have the opportunity to learn to critically evaluate different claims and differentiate reliable knowledge from hoaxes, conspiracy theories and 'alternative facts'.

One problem with this strong focus on facts is that students have little experience of how to apply their knowledge in everyday actions and situations, or how to use it to critically evaluate different political alternatives and form solid arguments for their own standpoint. Another problem is that it makes environmental and sustainability issues appear value-free and independent of ideology. In this perspective, these issues are not primarily a concern for democratic discussions, which means that little attention is given to discussions and students' deliberations in school. Consequently, fact-based teaching does not support students in their efforts to take a stand on environmental and sustainability issues and does not really prepare them for participation in democratic discussions.

Even more problematic is the companion meaning of this idea of facts and science as the ultimate problem-solver. This approach indicates a specific social order: experts define and instruct and the general public are informed. That is, the fact-based tradition may privilege a view of democracy that is called *technocracy* – a society run by experts.

When 'real' (scientific) knowledge is taken as a guideline for how we should relate to the natural world and social development, we also identify who has the right to speak about any related topics in classrooms, i.e. those who have access to 'real' knowledge. Access is relative, which implies that the person who has access to a large amount of 'real' knowledge has more authority to comment on which moral standpoints are acceptable than does someone with restricted access. In school, this means that it is the teacher and those students who are able to quickly assimilate 'real' knowledge who are accredited with positions of authority in the classroom. This puts limitations on students' rights to discuss and present their own opinions and ideas, as it is the teacher who is automatically appointed as 'the one who knows best'.

Normative – effective change but democratically questionable

The normative tradition presents sustainability problems as a question of attitudes and aims to encourage students to adopt a specific standpoint and take moral responsibility in relation to these problems.

The strength of the normative tradition is that it can be a very effective way of promoting individual change. Normative teaching can often create a social environment with a common cause. Combined with a space for emotions, this can make students very committed to environmental and sustainability issues. This commitment is not just a belief, but is properly based on scientific facts.

However, environmental and sustainability issues are becoming increasingly complex and have numerous dimensions (see chapter 2). One of the problems with this tradition is the difficulty of knowing which sustainability values, norms and perspectives to promote in education. How do we know which norms are right and which values are good? How can teachers be sure about their judgements? How do teachers know that the chosen version of sustainable development is the best one? It is one thing to take a personal stand and another to teach others (students in compulsory education) how to think and act.

When the good values and the correct norms are taken for granted, there is a risk that discussions exploring alternative options will be regarded as superfluous. If students are not part of the processes in which certain values are found to be most sustainable, it will be difficult for them to question the version of sustainable development that is being taught. When the focus is on encouraging a certain behavioural change, learning processes in which the students are given the opportunity to formulate their own views based on the available knowledge will tend to be neglected. Thus, in this tradition, the focus is on the socialisation of predetermined norms and values, rather than on the process of subjectification, where the individual student can create him- or herself as an autonomous political and moral person and find new ways of dealing with sustainability challenges (see chapter 4).

This tradition becomes particularly problematic when related to the democratic premise above, because focusing on one particular version of sustainable development automatically excludes other alternatives (see also chapter 15). If this teaching is successful, we end up with one single idea about sustainability. It thus limits the very basis of democracy, namely, pluralism – the diversity of ideas about what constitutes a good future society.

Further, in the case of participation in the classroom, the teacher adopts the role as 'the one who knows best'. This is similar to the fact-based tradition, although here the teacher is not only a scientific authority, but also a moral authority – 'the one who has good values and the right norms'. This distribution of authority in the classroom has crucial companion meanings about the role of citizens in society: They are the ones who obey the decisions made by politicians or other people in positions of power. Here, people in positions of authority are given *carte blanche* to decide what are acceptable/unacceptable attitudes and behaviour. This is not in line with democratic principles. A democracy allows each individual, based on knowledge, experience and values, to form their own understanding and develop a line of reasoning that supports their convictions. The normative tradition runs the risk of turning education into a political tool to create a specific predetermined society. This means that there is a danger that education will lose its emancipatory potential and that its democratic obligation will be violated.

Pluralistic – embracing democracy but a challenge to create a commitment

Like the normative tradition, the pluralistic tradition highlights the value dimension of sustainability issues. However, unlike the normative tradition, the pluralistic tradition aims to make students aware of the different perspectives and interests circulating in the sustainability debate. This does not imply that it is reasonable to question, for example, the overwhelming and carefully validated scientific evidence that climate change is induced by human activities. Rather, it means recognising that there are different opinions about how to best mitigate climate change, how quickly and at what cost – and that these alternatives are often conflicting.

It is also vital that students' own experiences and opinions are expressed in the classroom. It is in these discussions that the different standpoints and their implications are critically reviewed and valued. In this way, students are given the opportunity to develop their democratic action competence and learn how to take a stand, how to argue for their standpoint, and how to listen to other people's standpoints and take them seriously at the same time as critically reflecting on them. A particular strength of this approach is that students' standpoints become their *own* standpoints that are thoroughly anchored and hopefully durable. The pluralistic tradition accordingly turns the classroom into a democratic arena for negotiations about how to realise a sustainable future. This also means that the approach makes room for new ideas to develop.

The critique that can be directed towards the pluralistic tradition is that the democratic process takes up too much time and that the outcome of these processes is unpredictable. It is of course more difficult for students to take a stand on sustainability issues when there are good arguments and supporting facts for several different paths to a sustainable future. There might also be the risk of too much emphasis being placed on discussions at the expense of concrete sustainable actions. Advocates of the normative tradition claim that the environmental crisis is so severe that we need more effective methods for change. Another critique is that the pluralistic tradition implies a *relativism*: If we strive to illuminate different opinions about ethical and political issues in educational practice, would this not be interpreted as all alternative actions being equally right and all values equally good – that anything goes?

From a philosophical perspective, this critique is (only) relevant if we believe that universal values are based on something outside human practice. However, pragmatist philosophers like Richard Rorty (1980) have pointed to the difficulty of theoretically deciding whether our beliefs correspond to any such external foundation. It seems that it would require access to an unclouded picture of both our own beliefs and the external referent, or, in other words, that we occupy a position outside our language, culture and life (Englund et al. 2008; Öhman 2008; see also chapter 3). If we accept that values and the relation between facts and values is something that we establish among us human beings, then the question of relativism is rather a question of if the students understand the seriousness of sustainability problems, if they find them relevant and if they get emotionally

Table 5.1 Main characteristics of the three selective tradition of ESE

Tradition of ESE	Fact-based	Normative	Pluralistic
Perspective on sustainability problems	Sustainability problems are knowledge-based and are resolved by means of research and information	Sustainability problems are moral which can be resolved by exerting an influence on people's attitudes and behaviour	Sustainability problems are political which should be dealt with democratically
The cause of sustainability problems	An unforeseen result of production and resource exploitation in society	A conflict between society and the laws of nature	Conflicts between humans' wide range of achievement goals
Main method of teaching	Factual information from teacher to student	Transferring sustainable values in student active exercises	Critical discussions based on a number of alternatives
The purpose of ESE	Students receive knowledge of sustainability problems by learning of scientific facts	Students adopt sustainable attitudes and behaviour	Students develop their ability to critically evaluate and take a stand in sustainability issues
The aim of ESE	Citizens who have enough information to judge between different political alternatives in sustainability issues	Committed citizens who accept and approve of the necessary changes in order to develop a sustainable society	Citizens who are competent to engage in the democratic debate and practices that concern a sustainable future
Fact-value focus and relation	Facts	Facts → Values	↕ Facts ↔ Values ↕
The democratic process in relation to education	After	Before	In
Strengths and weaknesses	Based on reliable knowledge Omits the value dimension of sustainability issues	Effective for individual change Violates the democratic and emancipatory purpose of education	Supports democratic competence Time-consuming and a challenge to create a commitment

concerned. This is a didactical problem that has to be handled by the teacher: to highlight diversity, support free will formation, but at the same time encourage a strong commitment. If we succeed, then 'one cannot find anybody who says that two incompatible opinions on an important topic are equally good' (Rorty 1982/2003, 166).

Summary

By summarising the three selective traditions of ESE in table form, we are able to identify a number of significant differences between them (Table 5.1). If the fact-based tradition concentrates on *results* in the form of learning specific types of curriculum-based material, and the normative tradition focuses on *effects* in the form of sustainable attitudes and behavioural patterns, then the pluralist tradition is more concerned with being a catalyst of *processes*.

The traditions demonstrate three different ways of handling the fact-value relation in educational practice: (1) leaving values out of the rational discussion about sustainability issues, (2) deriving sustainable values from scientific knowledge and using education to transfer these values and (3) using pluralistic conversation to justify moral judgements and establish the relation between facts and values. These three different ways of handling the fact-value relation are related to three different perspectives on the democratic role of education: (1) to provide objective facts as basis for opinion-making, (2) to coordinate public will with political consensus and (3) to be an arena for democratic communication between autonomous subjects. The different democratic perspectives also imply that there is a difference in the view of where the democratic process is situated. In the fact-based tradition, the democratic process comes *after* education – education prepares students for participation in the democratic debate by providing them with essential knowledge. In the normative tradition, the democratic process comes *before* education – the political debate about which values and knowledge future developments should rest on, and consequently should be taught in school, precedes education. Finally, in the pluralistic tradition, the democratic process is situated *in* education – that a critical discussion about different alternatives and their consequences is an essential part of education itself. In this way, the traditions offer different conditions for the students' constitution as democratic citizens.

References

Andersson, K. (2017) "Starting the pluralistic tradition of teaching? Effects of Education for Sustainable Development (ESD) on pre-service teachers' views on teaching about sustainable development". *Environmental Education Research*, 23(3): 436–449.

Borg, C., Gericke, N., Höglund, H.-O. and Bergman, E. (2012) "The barriers encountered by teachers implementing education for sustainable development: Discipline bound differences and teaching traditions". *Research in Science & Technological Education*, 30(2): 185–207.

Borg, C., Gericke, N., Höglund, H.-O. and Bergman, E. (2014) "Subject- and experience-bound differences in teachers' conceptual understanding of sustainable development". *Environmental Education Research*, 20(4): 526–551.

Englund, T., Öhman, J. and Östman, L. (2008) "Deliberative communication for sustainability? A Habermas-inspired pluralistic approach". In Gough, S. and Stables, A. eds., *Sustainability and Security Within Liberal Societies. Learning to Live with the Future.* Routledge Studies in Social and Political Thought 58. Routledge, London, 29–48.

Öhman, J. (2004) "Moral perspectives in selective traditions of environmental education – conditions for environmental moral meaning-making and students' constitution as democratic citizens". In Wickenberg, P. et al. eds., *Learning to Change Our World? Swedish Research on Education & Sustainable Development*. Studentlitteratur, Lund, 21–32.

Öhman, J. (2008) "Environmental ethics and democratic responsibility: A pluralistic approach to ESD". In Öhman, J. ed., *Values and Democracy in Education for Sustainable Development: Contributions from Swedish Research*. Liber, Malmö, 17–32.

Östman, L. Svanberg, S. and Aaro Östman, E. (2013) *From Vision to Lesson: Education for Sustainable Development in Practise*. WWF, Stockholm.

Rorty, R. (1980) *Philosophy and the Mirror of Nature*. Princeton University Press, Oxford.

Rorty, R. (1982/2003) *Consequences of Pragmatism (Essays: 1972–1980)*. University of Minnesota Press, Minneapolis.

Rudsberg, K. and Öhman, J. (2010) "Pluralism in practice: Experiences from Swedish evaluation, school development and research". *Environmental Education Research*, 16(1): 115–131.

Sandell, K. Öhman, J. and Östman, L. (2005) *Education for Sustainable Development: Nature, School and Democracy*. Studentlitteratur, Lund.

Sund, P. and Wickman, P.-O. (2011) "Socialization content in schools and education for sustainable development I. A study of teachers' selective traditions". *Environmental Education Research*, 17(5): 599–624.

Swedish National Agency for Education. (2001) *Miljöundervisning och utbildning för hållbar utveckling i svensk skola (Environmental education and education for sustainable development in Swedish schools)*. Report 00:3041.

Williams, R.(1973) "Base and superstructure in Marxist cultural theory". *New Left Review*, 82: 3–16.

6 The ethical tendency typology

Ethical and moral situations in environmental and sustainability education

Johan Öhman and Leif Östman

Introduction

The ethical dimension of sustainability issues is about the good values that people find desirable and the right actions that reflect these values. It relates to questions such as who we ought to take into account in our strivings for a sustainable future and to what extent, whether we are only responsible for sustainable development in our part of the world or in the world as a whole, whether everybody has equal rights to the same welfare, whether future generations should have the right to the same welfare as us, how many generations we should be concerned about, whether future generations should have the right to experience wilderness and biological diversity, whether sustainable development concerns other species and whether animals and plants have the right to a secure future.

In this chapter, we present and describe a typology of the different ways in which morals and ethics can appear in educational practice. We will use *ethical tendency* as an umbrella term for all the different situations in which people express what they find to be fair, unfair, self-sacrificing, insulting, selfish, unselfish, greedy, generous, honest, dishonest, just, unjust etc. It is by participating in such situations that students can learn to communicate the right and the good and grow as moral subjects. But these situations can be very different in character, implying that the conditions of learning that prevail in these situations are diverse. It is therefore important that teachers are able to recognise these different situations so that they can adjust their teaching practices to the specific conditions of the event and support students in their ethical and moral growth in an appropriate way.

The typology is based on empirical analyses of video recordings of lessons, excursions and group discussions in diverse educational settings (Öhman and Östman 2007, 2008; Andersson and Öhman 2015) and on analyses of historical texts (Öhman 2006). We will use authentic examples from educational practice to describe the different situations and explain their differences. In the final section, we discuss the typology as a didactic tool for teachers to use when facilitating students' learning and guiding their moral growth in relation to environmental and sustainability issues. More specifically, the typology is related to a pluralistic approach to ESE (see chapter 5). Here we suggest principles for how to deal with the ethical tendency in a democratically responsible way and how to promote students' critical thinking and democratic action competence.

The typology builds on pragmatist philosophy and John Dewey's idea of morality as intimately associated with how we coordinate our actions with other people and our environment at a practical and social level (see chapter 7). The chapter is also inspired by the renowned philosopher Ludwig Wittgenstein's idea of ethics as a human tendency (1993) and his way of using reminders from real life to clarify philosophical questions (1953/1997, 1969/1997).

Ethical and moral actions

The question we would initially like to address is what makes us recognise an action as ethical or moral. What are the characteristic features of an everyday communication that concerns the ethical tendency? In order to clarify this question, we turn to a concrete example from educational practice as a reminder of the specific function of ethical and moral actions (the example is collected from Öhman and Östman 2008). In this secondary school lesson, the teacher initiates a discussion by announcing that there are plans to build a nuclear power plant close to the students' hometown. This announcement is followed by an intensive discussion amongst the students. Initially, the discussion mostly deals with the benefits and problems of nuclear power as compared to the usage of other energy resources. One boy, Peter, argues that nuclear power is preferable because 'electricity produced in wind power stations is more expensive than electricity produced in nuclear power plants'. Wind power is here judged against the purposes of power production in a modern society, that is, to produce power in an efficient and competitive way. Peter's claim has the function of a statement of fact. Usually we don't consider such judgements as moral or ethical. However, in response to Peter, Eva introduces a different reason for her particular stand:

Eva: But then wind power is much more environmentally friendly. Everybody says that you have to think of the environment and that it will be ruined if we don't improve.
Britt: Why should we destroy it even more then?
Adam: We will not be alive to see it.
Eva: But others will be living then, won't they?
Adam: It is a pity for them then.
Eva: Yes!

When Eva claims that we have a responsibility towards future generations, this judgement is not specifically connected to the purposes of power production. Eva rather expresses her judgement as if this is something that *ought* to be a concern for all people in every situation. Thus, we tend to express our ethical ideas and moral attitudes as *absolute value judgements*, that is, we communicate our beliefs in terms of *universally* good values and correct ways of acting. An absolute value is not something that you can confirm by relating to certain facts, but is something you simply believe in and is of deep existential concern. In everyday interactions, absolute value judgements have a specific function – an ethical or moral function. Different expressions of absolute value judgements can thus be regarded as a manifestation of the ethical tendency in human lives.

Ethical and moral expressions are not connected to the use of specific words. It is rather the case that words, facial expressions and gestures acquire meaning in relation to the certain circumstances of specific events. Thus, when learning to communicate the ethical tendency, we learn the meanings of different expressions in situations in which ethical and moral issues are at stake. In this way we can learn that even a frown can mean, 'don't do that, your behaviour is unjust!'

It is important to note that we generally can understand other people's ethical ideas and moral attitudes without sharing the same opinion. We understand them because we have had comparable experiences in similar situations, although this does not necessarily mean that our value judgements are the same. Hence, it is important to differentiate between a shared way of communicating ethics and morals and the ethical and moral opinions that different people hold.

The ethical tendency

In the following, we present the ethical tendency typology. The typology aims to serve as a didactic tool that we as teachers can use to distinguish between the various ways in which the ethical tendency can appear. The typology makes distinctions between three different kinds of ways in which attitudes to the right and the good are communicated: *moral reactions, norms for correct behaviour* and *ethical reflections*. Each of these situations is illustrated by real examples collected from educational practice.

Moral reactions

The term *moral reaction* is used as a reminder of those situations when we, without any previous consideration or reflection, take spontaneous responsibility for another being. For instance, this can happen when we see someone or something being treated badly, or when we spontaneously act in order to save someone or something that is in need. We can say that these are the situations in which we are *personally* affected and experience a feeling of care that reaches deep within us and that is often, even bodily, like a 'gut reaction'.

The following illustration is from an Outdoor Education Centre where children aged between eight and nine are catching animals in a small lake (the example is collected from Andersson and Öhman 2015). When a salamander that a group of children have captured in a tray suddenly seems to be dying, they get very upset and quickly try to put it back in the water:

Olle: Hold on now; hold on now [shouting].
Max: Everybody.
Linus: No wait, you too [shouting].

Max: Carefully [expressive voice].
Olle: Bye bye [they throw the salamander back into the water].
Max: Is he out?
Olle: Yes.
Linus: I can't see it.
Olle: There [points into the water].
Linus: No, it's not alive [sad voice].
Olle: Yes, it's alright, it's alright, it's swimming [shouting].
Max: No.
Olle: It is.
Olle: Yes it's swimming.
Pontus: This is almost cruelty to animals [serious voice].
Pontus: He did this [waves his hand in the air]; he is very scared [the salamander].
Olle: It'll survive.
Teacher: Is it alive? [Most of the children involved shout for joy.]

In this event, the children demonstrate strong emotional feelings of care and responsibility for the salamander and are very concerned about whether the creature will survive. The children react spontaneously without any observable consideration. The communication between the children is characterised by engaged exclamations, rather than reason or argument.

What we would thus like to highlight here are the unpremeditated reactions where we take absolute responsibility for someone or something – reactions that are not consciously forced or evoked by rational argument. If directly after a moral reaction we are asked why we reacted the way we did, we can normally only explain our behaviour in emotional terms: that we felt care, shame, agony, gratitude or guilt. Moral reactions cannot in any simple sense be perceived as being controlled by will; they are 'non-intentional'. Accordingly, personal moral reactions to our own and other human actions cannot be appropriately recognised as rational considerations or as following certain prescriptive rules.

Often, these reactions arise immediately and spontaneously in the actual situation. This is the case with the children in the example above. Another example can be when a person dives into freezing cold water to save a puppy from drowning. But we can also experience a moral reaction long after an event, such as when we suddenly realise that we have done something completely unfair.

Moral reactions do not necessarily come about when things happen before our very eyes. It is also possible to have such reactions when listening to a recital, watching a film or reading a book. This has important didactic implications, as it means that we as teachers can create situations that may evoke moral reactions that can be used as a starting point for an ethical inquiry.

Norms for correct behaviour

In our typology, *norms for correct behaviour* are a second way in which we experience opinions about the right and wrong way of acting. These are the situations in which the ethical tendency appears as *social rules* that are communicated to us by the way that authorities and peers actively respond to our actions. We could say that these rules prescribe the morally correct way of acting in certain kinds of situations, where we are expected to relate our actions to the social convention. In this way, the rules indicate a common opinion on how we *should* act.

The following event took place on the same day and at the same Outdoor Education Centre as the earlier example illustrating moral reactions (Andersson and Öhman 2015). At the beginning of the day, the outdoor teacher gave instructions on how to collect and treat the small animals:

Teacher:	Today it is also important to remember that we are just borrowing the animals for a while, in order to find a few things out about the animals. Afterwards all the animals will be put back into the water. They are just living creatures. Frogs and toads are a little bit special; we were just talking about toads. What do we need to think about when it comes to toads and frogs? Do you know?
Steve:	They need air.
Teacher:	They need air. Yes exactly, they actually need water and air. Is there anything else we need to think about regarding toads and frogs? Can you take them to school?
All at once:	No.
Teacher:	No, they are actually protected. What does protected mean? Someone from another class said that they were wanted. But they are protected.
Julia:	Er, well, that there aren't very many of them.
Teacher:	And that means that we mustn't move them away from here, but that they have to remain in this very place.

Here the teacher gives an account of what the rules are in this specific situation, namely that they are not allowed to remove frogs and toads from that particular place. The teacher informs the children about the common agreement for how to treat these creatures. The children are expected to act and take responsibility in relation to the teacher's demands.

This illustration points to an important difference between following a social norm and reacting morally. We follow norms because we *know we should* act in a specific way to fulfil the expectations of our fellow beings. Norms can therefore be formulated in terms of rules for the expected and accepted way to act. The existence of social norms for correct behaviour can be understood from the fact that our actions have consequences for the lives of our fellow beings. Our social environment therefore places certain demands on our actions. When we interact with other people, we learn what is expected of us by the ways in which authorities and peers actively respond to our actions by encouraging, condemning, confirming and questioning our actions. Thus, norms are the result of human beings living together, rather than ideals slotted into the lives of human beings (see Dewey 1922/1988 and chapter 7). When learning norms for correct behaviour we learn what actions are seen as ethical and moral, and we also learn what the right actions are.

The validity of the norms is generally linked to the doings of a particular community. Throughout our lives, we are part of many different communities, and thus experience different norms concerning the correct behaviour. For instance, when participating in an environmental NGO (non-governmental organisation), the norm might be that travelling by air should be avoided, while charter vacations are taken for granted at the workplace. When we grow up, we gradually learn to separate the norms of different activities and communities. Once we have learned the norms of a community, we generally take them for granted and do not have to consciously recall them every time we are about to act. In this way, the following of norms forms a social *habit*.

However, occasionally we distance ourselves from the norms for correct behaviour in a particular community and start a critical inquiry as to why a certain way of acting is considered right and certain values are seen as good. This is the third situation in our typology of the ethical tendency – *ethical reflections*.

Ethical reflections

In situations in which we *ethically reflect* on good values and the right way of acting, we often try to find systematic and rational arguments for how to handle certain moral issues. When we communicate and discuss such reflections, we are usually not in an immediate situation of needing to decide how to act. These ethical discussions therefore normally concern the *general ethical principles* that human beings *ought* to follow. In educational contexts, ethical reflections often appear as different forms of exercises, where students are expected to take a stand on a particular ethical issue and explain and defend their standpoints (see examples of values-clarification exercises in chapters 18 and 19).

An observed lesson in an upper secondary school contained such an exercise, in which the teacher made statements of environmentally ethical significance (the example is collected from Öhman and Östman 2007). Statements made by the teacher included: 'Man and animals are of equal value; it is always wrong to kill an animal!', 'No life must be sacrificed as a result of environmental pollution!' and 'Everybody has the right not to sort their garbage!' When formulating these statements, the teacher uses what could be called an *ethical language-game* (see Öhman and Östman 2008). A characteristic of these statements is that they do not refer to the moral reactions of any specific human being or the norms of any particular community. Instead, the statements are formulated as if they were valid for everyone in every activity, without any concern for the circumstances. This *ethical language-game* becomes apparent in the way the teacher uses formulations like 'It is *always* wrong to kill an animal!', 'No life must be sacrificed' and 'Everybody has the right'.

The students then took stands in relation to each statement and, in the ensuing discussion, defended their individual standpoints. In the discussion, one student responded to the teacher's statement that it is 'always wrong to kill an animal' in the following way:

Eric: The wrong thing is turning it into mass production; I mean to produce animals just to slaughter them. I can't change seats because I eat meat, and then I would be contradicting myself but I can understand the reasoning anyway, because I think it is a bit wrong to produce animals for this purpose. I can understand only killing animals in the wild, but there would not be enough animals for that. Because we want meat to eat, we have created some sort of mass production, so maybe it is wrong to do that, then maybe we should discuss in what way we treat animals. And there I think it is important, 'cause if you in some way declared equality between man and animals then you will come to the kingdom of plants at the next level, 'cause there are some who claim that they are just as equal as man and you should only eat windfalls or just eat a certain part. In this way it would always lead to a further debate. I think that animals hold the same value as man anyway, so I can agree on the first. But I will never stop eating meat so therefore I must agree on that we have the right to, right and right, but we do it because of . . .

This is an example of an *ethical reflection*, that is, a rational and systematic reflection on the reasons for our moral actions; a general enquiry into what is 'good' and 'right'. Thus, we can understand ethical reflections as a rational insight into moral issues. Such reflections are usually made at a distance from situations in which human beings actually have to perform the actions that they consider as correct. Moral issues and dilemmas are treated as disconnected from the particular circumstances that prevail in the lives of individual human beings. Sometimes the ethical reflection can start from a concrete example, but from that we strive to abstract general rules, norms or principles. In this way, ethical reflections can be seen as a way of justifying and forming *arguments* for our moral actions.

For a summary of the features that characterise the different situations of the ethical tendency typology see Table 6.1.

Table 6.1 Features that characterise the different situations of the ethical tendency typology

Moral reactions	Norms for correct behaviour	Ethical reflections
A spontaneous *feeling of care* or *disgust* Situations that affect you *personally* A kind of 'gut reaction' Not preceded by a rational consideration	Actions related to *collective rules* for correct behaviour – a moral social habit A kind of knowledge – something you learn from the response of others Connected to a specific *activity and community* To follow norms is necessary for being an *accepted member* of the group	Reflections on *the reasons for and consequences of our moral actions* *How we ought to act* in terms of general ethical principles A *distant* perspective: enquiries into what is 'good' and 'right' separated from the moral situation Rational and systematic way of dealing with moral problems

Didactical implications in the light of pluralism

In chapter 5, we argued in favour of a pluralistic approach to ESE involving diverse interests and perspectives, free opinion-making and democratic action competence. In this final section, we discuss the *didactical implications* of the suggested ethical tendency typology in the light of this approach. In relation to the three situations, teachers can make certain *ethical moves*, i.e. teaching actions that guide students' ethical and moral learning and promote their growth as moral subjects (see further about ethical moves in chapter 12). Which ethical moves would be suitable for a teacher who wishes to teach in line with the pluralistic tradition?

First of all, from a pluralistic perspective it is important that students encounter a *variety* of situations in which different expressions of the right and the good can be connected to the specific circumstances of these events. In this way, opportunities are created for students to articulate their moral reactions, ethical opinions and beliefs, which allows them to increase their sensitivity to the subtle nuances of language when it comes to communicating ethics and morals. However, this communicative learning is not automatically accompanied by the learning of a specific opinion about the good and the right, but it simply means that students learn a common way of understanding different expressions of the good and the right. Thus, increasing their communicative ability in this way does not mean that ethical differences are eliminated.

When it comes to *moral reactions*, it seems reasonable for teachers to take such reactions seriously and allow for a wide range of reactions. To systematically deny, neglect or argue against students' moral reactions, or their experiences of such reactions, would be to deprive them of their personal experience. Strong and perhaps diverse moral reactions in a class are often a fruitful starting point for an engaged discussion about ethical issues (see chapter 7). To systematically try to inculcate a certain way of reacting would, on the other hand, intrude on the personal sphere of the ethical tendency and expose ESE to the dangers of indoctrination. As teachers, we should therefore strive to make ethical moves that enable students to *express* and *share* their experiences of moral reactions. We should also give them opportunities to learn to respect the deeply personal, moral emotions that people show in different situations, even though these emotions may not always be possible to explain or defend by rational argument. By sharing narratives of situations with moral reactions, without moralising or attempting to convince anyone, students are offered opportunities to expand their awareness of different moral reactions and their ability to understand the various existential questions. Learning to take moral reactions into consideration is also a way of recognising that human actions are not entirely based on rational decision-making.

In order for activities to work, it is often a requirement for different opinions about right and wrong to be coordinated in common *norms* – also when it comes to actions related to environmental and sustainability issues. However, a straightforward transference of norms – such as which specific behaviour is regarded as being in line with sustainable development – would violate the participatory ambitions of the pluralistic perspective. It is therefore essential that we as teachers make moves that allow students to discuss the norms and the motives for the norms that are presented and that students are given an opportunity to critically reflect on and influence those norms. It is important to bear in mind, though, that learning to act in accordance with a norm does not necessarily mean that an individual is personally affected. This learning can rather be seen as the acquisition of social knowledge about how to behave in order to be an accepted member of a group. Another aspect of norms is that they are generally connected to the particular activity and community in which they have been learned. It is therefore an open question as to whether sustainability norms learned in school will influence the behaviour of individuals in their life outside its confines.

When it comes to *ethical reflections*, from a pluralistic perspective it would be problematic if we in our teaching practices stressed that *certain* ethical principles are more correct than others in our strivings for sustainable development. We would then limit the ethical diversity and ESE would narrow, rather than broaden, future possibilities. It is also important to bear in mind that most of our moral dilemmas are not solved by intellectual processes of rational consideration, where we can choose between ethical principles. If we strictly followed ethical principles in life, no dilemmas would even occur. But this would make us dogmatic and insensitive to the particular circumstances of the different moral situations we experience in life (see further chapter 7). The solution to a dilemma is often something that we discover as we live through it. This means that it is not possible to predict the extent to which the learning of ethical principles will influence the more crucial existential decisions related to environmental and sustainability issues that manifest themselves in students' lives.

On the other hand, working with ethical reflections opens up many possibilities. Exercises containing ethical reflections, where students evaluate different alternatives, formulate valid arguments for their standpoints, consider other people's arguments, learn more about their own and others' emotional reactions, etc. are very much in line with the pluralistic approach (see chapter 14 on argumentative discussions). It is reasonable to assume that in such exercises students will experience different forms of ethical reasoning and learn to relate critically both to their own behaviour and the norms they experience in school activities and in society in general. An important aspect of ethical discussions is also that ethical theory can be introduced, which can function as a way of transforming moral issues from the private to the public (see chapter 7). A recurrent integration of ethical reflections in educational practice can accordingly be seen as an important way of strengthening students' ability to participate in democratic conversations.

References

Andersson, K. and Öhman, J. (2015) "Moral relations in encounters with nature". *Journal of Adventure Education and Outdoor Learning*, 15(4): 310–329.

Dewey, J. (1922/1988) "Human nature and conduct". In Boydston, J.A. ed., *The Middle Works, 1899–1924, Volume 14: 1922*. Southern Illinois University Press, Carbondale and Edwardsville, 1–227.

Öhman, J. (2006) "Pluralism and criticism in environmental and sustainable education". *Environmental Education Research*, 12(2): 149–163.

Öhman, J. and Östman, L. (2007) "Continuity and change in moral meaning-making – A transactional approach". *Journal of Moral Education*, 36(2): 151–168.

Öhman, J. and Östman, L. (2008) "Clarifying the ethical tendency in education for sustainable development practice: A Wittgenstein-inspired approach". *Canadian Journal of Environmental Education*, 13(1): 57–72.

Wittgenstein, L. (1953/1997) *Philosophical Investigations*. Blackwell, Oxford.

Wittgenstein, L. (1969/1997) *On Certainty*. Blackwell, Oxford.

Wittgenstein, L. (1993) "A lecture on ethics". In Klagge, J.C. and Nordmann, A. eds., *Philosophical Occasions 1912–1951*. Hackett, Indianapolis and Cambridge, 37–44.

7 A pragmatist perspective on value education

Johan Öhman and David O. Kronlid

Introduction

Anna: We are so heavily dependent on each other.

Carl: Yes exactly!

Anna: We can live as we did before. Though I think that . . . we will have to change our lifestyle. Either we do it now and . . . preventing it from becoming a disaster like . . .

Beata: The thing is that we are stuck in this way of living, so we have kind of grown into it, so you are grown up with it and you are used to it. I am used to taking the car to different places et cetera and consuming the way I do and it is difficult to change my lifestyle. But as we so clearly see the consequences . . . it's quite obvious that we really need to change ourselves . . .

Doris: And realise that it actually makes a difference what you decide . . . so how to live etc . . . even though I'm just one human being so . . .

This conversation took place between some secondary school students during a lesson on climate change (the conversation was collected from Kronlid and Öhman 2014). This little sequence contains a lot of ethical (i.e. theoretically relevant) and moral (i.e. practically relevant) questions: How should we relate to other people? What is a good life? What responsibility do I have for my actions? To whom do I have a responsibility? Ethical and moral situations such as this occur constantly in the classroom, in relationships between students, between students and teachers and in relation to a specific subject matter. This dimension of education is also emphasised in the curricula for the preschool, primary school, secondary school and higher education in many countries. Thus, the goal of all our school forms is not only that students will develop certain subject knowledge, but also that they will grow morally and embrace fundamental democratic and human rights, such as individual freedom, gender equity, solidarity and inviolability of human life. This ambition is also present in UNESCO's *Education for Sustainable Development Goals Learning Objectives*, where it is stated that 'normative competency' is a cross-cutting key competency for achieving all the Sustainable Development Goals (SDGs), and that this competence includes:

> the abilities to understand and reflect on the norms and values that underlie one's actions; and to negotiate sustainability values, principles, goals, and targets, in a context of conflicts of interests and trade-offs, uncertain knowledge and contradictions.
>
> (UNESCO 2017, 10, Box 1.1)

To facilitate learning for this normative competency is therefore a central didactic skill for us teachers in all subjects, stages and schools. In addition to general knowledge in didactics, this includes knowledge in ethical theory and environmental ethics (see also chapter 6 for a typology of the different ways ethics and morals can appear in educational practice). In the first part of this chapter, we present a pragmatist perspective on ethics and morals specifically based on the works of the American philosopher and educational reformer John Dewey (1859–1952). We believe that this perspective can be particularly helpful for teachers in their understanding of ethics and morals in educational practice. In the second part, we suggest and discuss two teaching principles based on the pragmatist perspective which provide guidelines for a value education that supports a moral learning in line with the normative competency necessary for achieving the UN's Sustainable Development Goals.

Dewey's pragmatist view of ethics and morals

We often think of morality as something that someone 'has' (or does not have). Such thinking suggests that morality is an inherent mental property that can then be *expressed* in our actions. This way of thinking raises a number of difficult philosophical questions, such as: What is the relation between morality and action? Is there morality when we do not act? If so, where and in which form?

Dewey's philosophical projects are largely about bridging dualisms between something internal (for example, our moral perception) and external (for example, the reality and our actions) and thus solving questions of the kind mentioned above. In terms of ethics, Dewey does this by focusing on *how* morality works, rather than *what* it should be. Dewey's ethics are thus descriptive rather than normative. That is, Dewey is concerned with how ethics work in action and practice, rather than stipulating prescriptive ethical principles. In his argument, he reminds us that we need to start with moral practice also when engaging in critical ethical reflections on how we think and act morally, rather than only engaging in abstract reflections on how we ought to act and think.

Dewey has a broad view of what moral acts are: 'Morals has to do with all activity into which alternative possibilities enter. For wherever they enter a difference between better and worse arises' (Dewey 1922/1988, 278). Thus, he views morality as existing in and through our actions in our daily lives. In certain situations, this becomes apparent. An example is the student conversation provided at the beginning of this chapter. This transcript shows how a discussion about climate change can suddenly raise important moral questions in relation to lifestyle choices.

Dewey believes that morality is intimately associated with how we coordinate our actions with other people at a practical social level: 'Morals is as much a matter of interaction of a person with his social environment as walking is an interaction of legs with a physical environment' (Dewey 1922/1988, 318). As our actions always have an impact on other people's lives, we are therefore constantly in a moral situation.

As morality occurs in interaction with our environment, its communicative character is central: morality takes shape when our actions affect other people and we experience the consequences of our actions through the environment's reactions. Dewey calls this interaction between our actions ('doing') and the response from our environment ('undergoing') 'experience'. 'Experience' does not mean passively receiving impressions, but it is the active interaction of individuals with their environment (see Garrison 2001). Furthermore, it means that the environment should not be seen as determined and given in advance – what the environment is for us is how it responds to our actions, which in turn depends on how we act. Our 'moral experiences' function in this way as a kind of mutual functional coordination with our social environment.

Coordination with our social environment allows us to develop patterns in our ways of acting – we establish habits that include norms for the right way to act. The norms make it possible for us to work in everyday life without having to reflect on what is morally better or worse every time we act – it is obvious to us how to act. However, the norms should not be seen as something to be mechanically followed, but as a repertoire of possible actions that are available to us in different situations. The more experience we have, the more purposefully or morally we can act. Dewey rejects the idea of a congenital moral conscience and instead believes that morality is contextual because it is the social environment in which we grow up that determines which moral judgements we develop.

However, our norms are not always adequate in all situations; sometimes we end up in situations in which we feel morally insecure. Such situations are not a simple choice between right and wrong but are *dilemmas*, where we have to choose between actions that from different perspectives can be considered good or less good (it could also be about choosing between two negative consequences). Such situations require us to *reflect* on our choices and consider the *moral consequences* they may have (intelligent actions). Dewey emphasises humans' moral imagination, i.e. that with the aid of our language we imagine different ways of acting and how these various actions favour or affect our fellow human beings (Fesmire 2003; see also Arneback 2014). Therefore, we do not need to perform an action to understand its potential moral consequences. When we reflect, we put action and consistency in relation to our previous experiences: 'Every experience influences in some degree the objective conditions under which further experiences are had' (Dewey 1938/1997, 37). This does not only mean that we recall previous experiences to deal with our moral dilemma, but also that these experiences are crucial for how we perceive the dilemma itself.

The different concrete situations we live through in our lives are both the means and ends of morality (see Pappas 2008). The situation is the means of morality because it is in the specific context of the situation that we can identify different ways of acting and find support and guidance for our moral choices. The situation is the end of morality because it is the consequences for those involved in the specific situation that determine what is better or worse. This contextual and consequential ethics means that Dewey rejects the idea that universal, normative, ethical principles can determine what the right choices are. Instead, Dewey believes that morality is continuously evolving, both for us as individuals and in society, in that we are constantly exposed to situations of insecurity and conflict where we are forced to reflect on our choices. Adhering to universal ethical principles would hinder this development, because we risk becoming insensitive to the circumstances of the particular situation.

Therefore, from a pragmatist perspective, moral questions can never be reduced to a collection of universal ethical principles and ideals that are applied to everyday situations from outside. At best, we can formulate ethical principles afterwards as a consequence of our experiences. In this way, Dewey believes that ethical principles can be seen as a 'condensed' result of the moral experiences that people have in their daily lives and that these are useful as resources when making choices. One example is how the principle of 'not lying' can serve as a guide in many contexts. However, it is also possible to imagine situations in which a lie would spare other people from unnecessary suffering and that we therefore find it morally acceptable to lie in certain circumstances.

Democracy – a moral ideal

We mentioned above that Dewey discussed above all the nature of morality, but he also argued for a number of moral ideals, and perhaps the most prominent ideal in his texts is *democracy*. For Dewey, democracy 'is more than a form of government: it is primarily a mode of associated living, a conjoint communicated experience' (Dewey 1916/1980, 87). This means that Dewey's democracy is *not* a struggle in which people relate to each other by emphasising and defending their predetermined positions but a way of life in which people with different experiences *create new possibilities* by influencing each other.

Moving from a view of democracy as a form of governance into a way of life makes democracy more profound than everybody's right to express themselves and other formal democratic rights (Pappas 2008). It's more about how we meet each other, what attitudes and approaches we have to other people and thus how we create equality in everyday life.

Democratic respect is not just about how we treat others ('doing') but is as much about how we receive and relate to other people ('undergoing'). A democratic attitude implies a sensitivity to people's different conditions – to appreciate other people as individuals who have grown in relation to their specific circumstances and therefore have partly unique opportunities to continue to grow. Thus, democracy can be seen as a communicative practice in the same way as ethics and education. According to Dewey, all these communicative practices should aim to create the conditions for new perspectives, thoughts and ideas to emerge and to be critically evaluated so that we can grow both as individuals and as culture and society.

Consequences for education

Dewey's democratic ideal has crucial implications for his view of school and education. He believed that school is one of the most important *democratic arenas* in our society (Dewey 1916/1980). The reason for this is that in school people meet with different experiences and backgrounds, such as class, religion and ethnicity, in common meaning-making processes. This means that students can learn a moral democratic ideal by experiencing it in their school education. It is about democracy as an approach that is integrated into teaching – to learn *in* and *through* democracy rather than about democracy. In this way, Dewey meant that one of the central functions of school is citizenship education and to lay the foundation for a coexistence in a pluralist society.

A central part of this citizenship education is to promote students' moral growth. Here Dewey's view of morality as always situated in the interface of person and context has important consequences. First, morals are not innate or fixed, but are something that we *learn*, and this is a continuous process throughout life. We learn by participating in moral situations that make us reflect on responsibilities and concerns that we have either previously taken for granted or communicated to others. It is thus an interactive, embodied and situated learning experience. In this way, we can gradually learn to be more sensitive to the specific circumstances that prevail in diverse moral situations and develop an intelligent sympathy. This makes the didactical aspect of ethics of great concern and is why it is important for teachers to reflect on the ethical and moral dimensions of their teaching practices. Teachers' didactic choices of content and methods always have moral implications, in that they predispose the students to certain actions:

> Perhaps the greatest of all pedagogical fallacies is the notion that a person learns only the particular thing he is studying at the time. Collateral learning in the way of formation of enduring attitudes, or likes and dislikes, may be and often is much more important than the spelling lesson or lesson in geography or history that is learned. For these attitudes fundamentally are what count in the future.
>
> (Dewey 1938/1997, 48)

What Dewey points to here is that we as teachers unavoidably and often unintentionally send messages of moral significance when we teach. In chapter 4, the values and norms that follow the learning of knowledge is described as companion meanings. The moral consequences of everyday teaching, therefore, will be important for teachers to consider in their didactic choices of content and working methods. The presence of companion meanings implies that students' value education should not (only) be a limited educational content, i.e. something that is involved in certain subjects during certain hours. Rather, it is about a recurrent raising of moral questions in all subjects, for example in relation to textbooks, websites and other teaching materials. It is also about paying attention to the moral questions raised by students in their daily work (as in the introductory example) and use these as opportunities for further ethical reflection and to discuss the moral consequences of potential actions. However, this does not exclude the need for specific education in ethics and morals.

Principles for teaching ethics and morals

In the following sections, we discuss how teachers can teach ethical and moral issues in a way that enables learning that is in line with UNESCO's key competence for dealing with complex and value-laden sustainability issues. We will do that by suggesting two simple principles for teaching morals and ethics based on Dewey's view on morality and his idea of democracy as moral ideal: (1) start in students' moral experiences of concrete cases and (2) introduce ethical theory and language. To follow these principles can be understood as a way for teachers to organise their 'ethical moves' systematically (see chapter 12).

Start in students' moral experiences of concrete cases

This principle means to start teaching in students' moral experiences that emerge in real-life, concrete cases of complex sustainability issues. This teaching principle directly connects to Dewey's view on morality as always situated in the interface of person and context.

The point to start in students' moral experiences is to make sure that the teaching is both *relevant* and *concrete*. It makes the teaching relevant because it is the students' experiences, rather than fictive ethical or philosophical examples, that govern the content of the teaching. Following Dewey, this ensures that value education is integrated with the students' life. This also means that the moral learning involves students' emotional responses to the problem at hand. Starting the teaching in real cases also makes the teaching concrete because cases are always time- and actor-bound.

The cases and students' experiences can be of different kinds. It can be in the form of previous experiences that students recall in relation to the teaching content of the specific lesson. Another example is when the teaching content evokes spontaneous moral responses as in the introductory example. But it can also be the case that the teacher deliberately wants to provoke a moral experience, for example, by showing a movie or reading a text which concerns the students and arouses their emotional responses.

To start in real experiences can sometimes be painful as it confronts our own shortcomings in relation to our moral ideals (see Baumann 1993). It is therefore often appealing to theorise the problem and turn the problem into a rational discussion of ethical principles, where the strength of different arguments is tested on purely logical and argumentative grounds. But if we as teachers move too quickly to the level of abstract principles, there is the risk that this makes the students escape from the moral problem and prevents them from having a personal experience of the problem. What we argue for is rather that we as teachers should help students to *live through* the problem.

This can be done through an ethical reflection that considers the concrete conditions for the well-being of those encountered in the specific case. It is important to allow students to communicate their (often emotional) moral responses openly, without the teacher imposing epistemological (they are incorrect), formal (they are badly formulated) or moral (they are immoral) norms. In order to help them to deepen their understanding of the context of the case, the students are encouraged to seek relevant knowledge (e.g. the historical background, scientific explanations and political and cultural conditions of the case). This knowledge stretches beyond strict ethical theoretical knowledge and demonstrates that moral philosophy and ethics, as educational content, are highly relevant to all school subjects.

An example of moral experience that can be the starting point of teaching is the moral concern students may experience in relation to the fact that affluent lifestyles cause climate change, which, in turn, has a negative effect on human and non-human well-being in the present, globally and intergenerationally. That is, students may feel trapped and path-dependent in a particular lifestyle that they know is one of the indirect causes of human and non-human suffering, while simultaneously cherishing that same lifestyle because it supports their own flourishing. Here, scientific knowledge about lifestyles and climate change can be connected to students' feelings and emotions in relation to the existential question of whose well-being to prioritise. Hence, connecting to concrete cases will force students to remain in the problem and work on possible ways of navigating the particular moral questions that emerge from their own experiences.

When sharing moral experiences in the classroom, it is important to focus on the *personal* rather than the *private*. By private, we mean that something only *concerns* the individual student while personal means that something *affects* and goes deep within the individual student. If a moral experience and reflection is only of private concern, it will soon run its course, and there is a risk that the learning process will not be as educative as it could be, because any reflexive process needs external input. Hence, the reflection process needs to be communicative and of moral relevance, that is, it needs to *mean* something to the students beyond intellectual excitement in order to generate the kind of normative competence that is emphasised in the UN's Sustainable Development Goals.

Introduce ethical theory and language

This principle means to connect students' moral experiences with ethical theory and an ethical language. In Dewey's terms, this is a way to encourage the students to go from primary moral experiences to a reflective moral meaning-making process and in this way enable intelligent moral actions and communication. It is about giving moral emotions and feelings a wider meaning which can be articulated and shared with others.

Whereas the principle of moral experience primarily ensures that the reflection process does not vanish into abstract reflection space by anchoring the reflexive process in particular *concrete* and *relevant* cases, the principle of ethical theory is necessary to ensure students' *progression* in the value education. Progression is enabled by using conceptual frameworks and typologies from ethical theory to further structure the content of students' moral experiences in the reflection process. Such reflections might involve, for example, the extent to which they find future generations, other species and geographically distant humans to be morally considerable. Central is also the question of the relative moral 'weight' morally considerable entities should have in cases of conflict, i.e. whose well-being should we prioritise and to what extent (see Goodpaster 1993 and Kronlid and Öhman 2013 for a framework concerning different possible environmental ethical positions on this issue).

The introduction and use of ethical theory offer students a partially new language by which they can relate to their own and others' morals. This language, which will be a mixture of students' personal expressions and theoretical expressions, constitutes an important part of the teaching in ethics education. As the students in this way use new theoretical concepts to discuss these issues, they gain access to new ways of reasoning, formulating arguments for their own position, considering the arguments of others and examining the validity of the arguments in a social context. In this way, students get the opportunity to develop their own intersubjective language with which they can orient themselves in the moral space and thereby understand their own and others' moral feelings and positions. This relates to Dewey's idea of moral meaning-making as a communicative practice and that sharing moral experience can create new insights beyond past and individual experiences.

Using ethical theories can thus be a way to create a conversation that challenges students' moral thoughts rather than just confirming or rejecting them. A didactic purpose of allowing the reflection to be guided by a query derived from ethical theory is that such an education takes the students' moral experiences seriously, without sacrificing the theoretical knowledge needed to develop knowledge in the field of ethics. Ethical theory can thus be a tool that helps students to structure their moral standpoints and supports their moral development.

Using ethical theory as a tool for students' moral development is also an alternative to the doubtful idea that the teacher should be the moral model that guides this development. Such an ideal model thinking runs the risk of reducing the possibilities of value education and depriving students' right to an education of good quality, regardless of the teacher's individual values and way of life. The main reason for this is that value education is a professional endeavour and should, just as any other subject or sub-subject in school, not be based on the character of the teacher. Why? First, because we teachers are human, and humans react differently to other people's moral responses and arguments. These responses are naturally linked to the teacher's own moral outlook on what is right and wrong. Hence, if teachers are to teach based on their own moral character or convictions, students who do not share these values are likely to be excluded from the reflection process simply because they are, measured against the teacher's moral outlook, (morally) wrong. Second, value education (and we would argue any education) is not primarily a moral task, but a pedagogical one. The teachers' main task is to ensure that the learning situation is as productive as possible for the student. Hence, basing value education on the teacher's moral convictions and outlooks, instead of enabling a critical open reflection process, will most likely reduce the possibility of establishing such educative moments.

Concluding remarks

In the Deweyan perspective, moral questions cannot be reduced to a collection of universal ethical principles that are applied to everyday life situations. Rather, ethical principles are formulated as a consequence of our experiences. In this way, Dewey believes that ethical principles can be seen as a 'condensed' result of everyday moral experiences and fundamentally a learning process. Applying the principle of moral experience and the principle of ethical theory as a basis for different ethical teaching moves is a way to lead students in this learning systematically (see chapter 12). The purpose is to offer students an opportunity to reflect on their own and others' moral experiences, formulate arguments for their own position, consider the arguments of others and test the validity of these arguments in their own social contexts. By encountering different moral responses and ways of reasoning ethically, it is possible for students to develop a critical approach to, and possibly re-evaluate, their own and others' moral responses, statements and actions in real-life situations. In addition, this will stimulate critical reflection on how the students' moral attitudes relate to generally accepted and emerging ethical norms of behaviour. This kind of value education can, we argue, enable the normative competence that, according to UNESCO and others, is a key competence for dealing with complex value-laden sustainability issues.

All teaching is governed by its purpose. So too is the teaching of ethics and morals, and it is the teacher's task to manage the tension between personal experiences, existing societal moral norms and values and ethical reflection. Just as Dewey wants to refrain from normative ethics by focusing on the *how* of morality, our aim with this chapter has been to give an example of how teachers can engage in value education in a way that does not deflect from the normative, but enables a space in which students can critically and systematically reflect on their own and others' moral experiences and values, how they argue and what they base their normative convictions on – and ultimately to help them navigate a highly complex moral space generated by sustainability issues. Doing this is much greater, although more difficult, than simply telling students how to act and think.

References

Arneback, E. (2014) "Moral imagination in education: A Deweyan proposal for teachers responding to hate speech". *Journal of Moral Education*, 43(3): 269–281.
Baumann, Z. (1993) *Postmodern Ethics*. Blackwell, Oxford.

Dewey, J. (1916/1980) "Democracy and education". In Boydston, J.A. ed., *The Middle Works, 1899–1924, Volume 2: 1925–1927*. Southern Illinois University Press, Carbondale and Edwardsville.

Dewey, J. (1922/1988) "Human nature and conduct". In Boydston, J.A. ed., *The Middle Works, 1899–1924, Volume 14: 1922*. Southern Illinois University Press, Carbondale and Edwardsville, 1–227.

Dewey, J. (1938/1997) *Experience and Education*. Touchstone, New York.

Fesmire, S. (2003) *John Dewey & Moral Imagination: Pragmatism in Ethics*. Indiana University Press, Bloomington.

Garrison, J. (2001) "An introduction to Dewey's theory of functional 'trans-action'. An alternative paradigm for activity theory". *Mind, Culture, and Activity*, 8(4): 275–296.

Goodpaster, K. (1993) "On being morally considerable". In Zimmerman, M. ed., *Environmental Philosophy: From Animal Rights to Radical Ecology*. Englewood Cliffs, Prentice Hall, 49–65.

Kronlid, D.O. and Öhman, J. (2013) "An environmental ethical conceptual framework for research on sustainability and environmental education". *Environmental Education Research*, 19(1): 21–44.

Kronlid, D.O. and Öhman, J. (2014) "Etik och moral som en del av utbildningens praktik" ("Ethics and morals as a part of educational practice"). In Jakobson, B. and Wickman, P.-O. eds., *Lärande i handling: en pragmatisk didaktik*. Lund, Studentlitteratur, 185–194.

Pappas, G.F. (2008) *John Dewey's Ethics: Democracy as Experience*. Indiana University Press, Bloomington.

UNESCO (Rieckmann, M.). (2017) *Education for Sustainable Development Goals: Learning Objectives* (https://unesdoc.unesco.org/ark:/48223/pf0000247444). Accessed 1 March 2019.

8 The political tendency typology

Different ways in which the political dimension of sustainability issues appears in educational practice

Michael Håkansson, Katrien Van Poeck and Leif Östman

Introduction

In this chapter, we present and describe a typology of different ways in which the political dimension of sustainability issues can appear in educational practice. This typology, that we have labelled the 'political tendency' (Håkansson et al. 2018), has been constructed in a similar way as the ethical tendency typology (Öhman and Östman 2008) presented in chapter 6 but here with a focus on identifying and classifying a diversity of situations in which the *political* dimension of sustainability concerns appears in environmental and sustainability education (ESE). Thus, instead of paying attention to instances of communicating the good values that people find desirable and the right actions that reflect these values (i.e. the ethical dimension – see chapter 6), the political tendency has been created by classifying a wide variety of situations in which people engage with the question *how to organise society*, acknowledging that this inevitably requires judgements, prioritisations and decision-making about different and competing alternatives. This political dimension of sustainable development thus concerns questions such as 'How shall we distribute the planet's scare resources among all citizens?', 'Which rules and laws can contribute to a more sustainable production and consumption of food, energy, etc.?', 'Who is responsible for an unsustainable situation?', 'How do we end extreme poverty and inequality?', 'How should we organise traffic in the city?', 'What is the right way of governing a transition towards a more sustainable world?' and 'Who shall have a say in this?' In sum, the political tendency is an umbrella term for all the different situations in which people express what they find good, right, just, sustainable, etc. ways to organise our society. It becomes visible in ESE when teachers and students, in their everyday practices, communicate opinions, judgements and prioritisations about how a good, sustainable society should take form, thereby encountering different visions and perspectives. These situations can be very different in character, and hence the conditions for learning that they bring about are also very diverse. In order to support students' learning in an appropriate way, it is important for us as teachers to be able to recognise these different situations. Then we can adjust our teaching practices to these specific conditions in relation to the specific purposes of our lessons and courses.

The political tendency typology is created through empirical analyses of several formal and non-formal ESE practices (Håkansson et al. 2018). Exploring these very diverse teaching and learning practices, we have identified and described a rich variety of practical manifestations of the political dimension of ESE. The different situations in the typology hence range across often-made theoretical distinctions such as that between 'politics' and 'the political' made by post-foundational political scholars such as Chantal Mouffe (2013). Whereas 'politics' refers to dealing with everyday political problems through strategies, actions, procedures, worldviews and institutions that seek to establish a certain hegemonic order in the organisation of society, 'the political' refers to the dimension of antagonism that is inherent in human relations and prevents a final closure of any given order. Mouffe emphasises that politics are always, inevitably conducted in conditions that are potentially conflictual and that every hegemonic order is contingent in the sense that there is no objective (moral, rational) ground that can serve as a foundation for solving political problems. Rather than starting from such a priori, theoretical constructs and distinctions, the political tendency typology has been created by identifying and ordering the plurality of observed *practical* manifestations of students and teachers engaging with the question of how to organise society, acknowledging that this inevitably requires judgements, prioritisations and decision-making about different and competing alternatives.

As argued, the political dimension takes diverse manifestations in concrete ESE practices, which have various impact on students' learning. We distinguish between 'political norms', 'political reflections', 'political deliberation' (sub-divided into 'normative deliberation', 'consensus-oriented deliberation' and 'conflict-oriented deliberation') and 'political moment'. Each of these categories are explained and illustrated in the subsequent sections. In the final section, then, we discuss the didactical implications of the typology. Many ESE researchers have emphasised the importance of creating spaces for students to express, explore and confront a plurality of dissonant and conflicting voices (e.g. Wals and Heymann 2004; Lundegård and Wickman 2007; Knutsson 2013; Sund and Öhman 2014; Van Poeck et al. 2016; Håkansson et al. 2018). As we will further elaborate below, the different categories in the political tendency typology bring about very diverse educational settings that vary largely as to the opportunities offered to students to learn to handle dissonant and conflicting voices.

Political norms

In our typology, *political norms* are a first way in which the political dimension of sustainability issues can appear in educational practice. These are the situations in which the political tendency appears as the communication of rules that prescribe the democratically correct way of acting in certain kinds of situations. This is a very particular way of addressing the question of how to organise social life, namely through an exclusive focus on the form and process of participation in democracy or, more precisely, on how to distribute voices and power within the classroom and the school as an institution.

The situation of political norms can be illustrated with an example of an eco-schools project at a primary school. The project wants to raise awareness of sustainable development through engaging students in designing actions and activities that make the campus and school environment more eco-friendly and sustainable. Thus, the aim is that students learn how to participate in decision-making about sustainability issues in a democratic way. Especially the activities of eco-school student council meetings exemplify this strong focus on training the students to participate in a democratic way in the organisation of the social life in the school. During one of these meetings, the teacher and a group of students discuss the organisation of a march for sustainable transportation in the school's neighbourhood. After giving each student the opportunity to raise ideas – thereby emphasising that 'All ideas are good, right?' – the teacher continues as follows:

Teacher: Okay. We heard some things now. [. . .] Good. My question for you is that you just think for a while and say, or think, what's for you, in your own view, uhm, the best thing to do Friday. Then we'll see what the majority decides and we can then maybe, if you don't fully agree, change it a bit. Okay. Is there already someone who says I want it like that? But we let the others have their say, right? Everybody gets the chance and then we see. I'll start with you and then follow the circle. [the teacher writes down all the students' responses]

Jürgen: The idea of using music instruments. And Tom's idea, to play the instruments together.

Tom: The same as Jürgen, yes, doing it together. But those who play the instruments have to be in the front of the march, with the flag.

Lotte: I would do the same as Jürgen and Tom.

Teacher: Are there any other kids who think like that? [no one responds – the teacher appoints another student]

Sam: I don't know yet.

Teacher: I will get back to you then.

Elke: I would do . . . all the kids who play an instrument, and then what Kobe said, that we'd drum on pans with wooden spoons to make noise.

Anouk: I don't know . . .

Teacher: Kobe?
        [Kobe shakes his head – another student does so too]

Hans: Also the same as Jürgen.

Jana: Also the same like Jürgen.

Sarah: I don't know.

Rune: The same as Jürgen.

Teacher: I have a lot of marks for Tom's idea and the idea of playing real instruments. Those of you who didn't know yet, can you agree with this idea?

Sam: Yes.

What this activity illustrates is a strong focus on the form and process of participation in decision-making regarding how to organise social life within and around the school. The teacher organises this in line with a specific political norm, i.e. in such a way that the different voices and thereby different alternatives will be heard. By communicating these norms (for example 'All ideas are good, right?'; 'what's for you, in your own view, the best thing to do?'; 'the majority decides and we can then maybe, if you don't fully agree, change it a bit'; 'we let the others have their say, right? Everybody gets the chance'), all students are included in the decision-making process and trained to participate in a democratic way in the sustainability governance at school.

Political reflections

The second category in the political tendency typology is *political reflections*. Here, contrary to situations of political norms, the focus is not on the form and process of participation in democratic decision-making but on creating new or refined knowledge and understandings of the political dimension of sustainability issues. The aim of teachers, then, is to encourage students to reflect rationally on how we organise social life as well as on the different, competing opinions, judgement and prioritisations regarding what are good, right, just, sustainable, etc. ways to organise our society. To engage in such reflections, students use prior knowledge and understandings.

The coordinator of the above described eco-schools project illustrates this focus on political reflections while explaining that one of the purposes of the project is to facilitate students' reflection on broad questions related to the sustainability problems they encounter:

> When a situation or problem occurs, which students are involved in from the start, like okay, what do we find here? Where does it come from? Why does this happen or not? . . . Is it a good or a bad thing? Are we satisfied that this happens or is it a problem? Why is it a problem? Who's the victim, right, or who benefits? Who suffers from it? And if they are motivated then to . . . what can we do about it? What would be possible solutions?

Questions such as 'is it a good or a bad thing?', 'who suffers from it and who benefits from it?' and 'What would be possible solutions?' open up a space for reflecting on different competing alternative answers on the question how to organise society as well as on the existence of conflicting interests behind prioritisations, solutions and decisions.

From the perspective of students, the result of such political reflections can look as follows. After a classroom discussion in a secondary school about how to reach a sustainable society when citizens are unwilling to fulfil that aim, the students were given a questionnaire and asked to write down their reflections on the discussion (see also later section, 'Consensus-oriented deliberation'). Students wrote, for example, 'After this kind of lesson you start to think of how many opinions that should be allowed in our society' or 'got a new way of seeing how different you can be seen on foremost political decision methods', which shows how they had reflected on whose opinions should count, and whose should not in decision-making processes. The cognitive activity of political reflection resulted in new insights and understandings but did not make the students take a personal stand and defend it. The latter is the major difference in relation to the practices of political deliberation that we present and discuss later.

Political deliberation

Political deliberation is a third category in the political tendency typology. Here, too, political reflections occur, but these reflections are made in an argumentative context, i.e. when students engage in an argumentation and defend their personal opinions, standpoints or preferred alternatives in light of a real or potential decision regarding societal issues. Political deliberation can take shape in three different forms: *normative deliberation, consensus-oriented deliberation* and *conflict-oriented deliberation*.

Normative deliberation

It is not unusual in an educational context that the expected outcome of a deliberation is already set by the teacher but not known in advance by the students; the agenda is implicit. Because of the pre-set outcome of the argumentation, i.e. its orientation towards an a priori specified result (in the form of a particular stand regarding how to organise society) as the only correct one, we call this category *normative deliberation*.

An example from a workshop on the ecological footprint in an adult education context illustrates how an educator steers the deliberation about how to reduce our environmental impact towards an a priori defined solution (see also chapter 13). After an explanation of the concept of ecological footprint (see chapter 1), small groups of participants discussed behaviour clues for reducing their ecological footprints. Subsequently, they presented their findings for the whole group.

Ellen:	One obvious improvement would be to bike or walk short distances.
Teacher:	What would you consider a short distance?
Ellen:	Going to the bakery, for instance.
Teacher:	How much is that in miles?
Ellen:	One and a half?
Teacher:	No, let me help you out: in fact, we should bike any distance under 3 miles.
Caroline:	Hello-o!! [laughter]
Teacher:	Why 3 miles? Because cars consume most over short distances.
Ellen:	Then we also had to say why we found it difficult. We found it can be time-consuming at times.
Teacher:	Remember the word I just used: planning?

After the participants' response to the educators open question what they consider a short distance, the educator disagrees with their answer and introduces a norm: 'we should bike any distance under 3 miles'. Thus, the question of how to organise society in view of more sustainable ecological footprints is answered in a normative way, stating that everyone should behave according to a certain, very specific standard. Throughout the activity, the teacher repeatedly intervened in similar ways (see also chapter 13). Thus, the frame of deliberation becomes normative: of all the different, competing perspectives on how to make society more sustainable, only one solution counted as valid. In that way, the normative variant of political deliberation differs from the two others described below.

Consensus-oriented deliberation

In a *consensus-orientated deliberation*, the deliberation is characterised by its orientation towards coming to a consensus that is not a priori set by the teacher. The students deliberate in the atmosphere that all divergence (opinions, judgements, prioritisations, etc.) can finally be resolved into one shared standpoint or decision.

Let us return to the classroom discussion in a secondary school about how to reach a sustainable society when citizens are unwilling to fulfil that aim (see political reflections). Before writing down their reflections, the students had been discussing this topic for about 30 minutes in two groups without the teacher being involved. Then, the teacher joins one of the groups and asks them to sum up their discussion:

Mark:	[inaudible] people stop whining and not go on nagging about everything.
Sven:	That was your suggestion.
Sophie:	[points at Mark] I agree on that.
Christine:	Freedom in every aspect is bothersome as hell.
Mark:	It was an alternative that we brought up. I didn't say we agree on that. It was an alternative that I brought up, it was a position that I pushed.
Amelia:	I agree with you [inaudible].
Mark:	Democratic reduction for the best of the public.
Christine:	And you know what that is?
Mark:	Yes. [long pronunciation]
Amelia:	We agree that maybe it was a bit too much freedom, thus in some cases . . .
Sven:	. . . Yes a bit too much freedom . . .
Amelia:	Little too much to choose from and a few too many rights in some cases. Cannot we agree that we thought that?
Teacher:	A few too many rights and a bit too much freedom?

The students' argumentation shows that the deliberation was directed towards reaching mutual agreement, particularly on the idea that there is too much freedom in the society, too much democracy. A very similar conclusion was expressed in the other small group and put forward in the subsequent whole-class discussion.

Conflict-oriented deliberation

In contrast to normative and consensus-oriented deliberation, what characterises *conflict-oriented deliberation* is that it opens up for the conflictual in the deliberation through having the students raise and defend opposing and contesting perspectives, without the aim that a consensus has to be reached.

The following event took place during a group of university students' excursion to an ecological farm (see also chapter 13). The farmer took time to elaborately express and explain his own points of view on why and how to run a farm sustainably and thereby repeatedly invited the students to say so if they would disagree with him. So doing, he urged the students to take a stand, to defend their personal position with arguments, to contest his and each other's points of view with counterarguments, etc. In the beginning they were a bit reluctant, but as the conversation continued more and more students voiced divergent opinions and got engaged in a conflict-oriented deliberation. For example, when the issue of agricultural subsidies was raised:

Farmer: I don't receive any subsidies. And I also think that it would be very good to say that we are putting an end to them.

Emma: But *you* also don't live from [agriculture]! [original emphasis – raises her voice]

Farmer: I *do* live from it. [original emphasis]

Emma: Oh, you said yourself that you don't pay yourself a wage! [raises her voice]

Farmer: Yes, but that's different. You don't need a wage to be able to live from it. I eat from it. That's a big difference. If you think I've got 2,000 euro on my account at the end of the month. I think I've got 900 euro or something like that on my account.

By taking a firm, personal stand, the farmer highlighted the existence of different and competing perspectives on how agriculture should be organised in our society. What he suggests – putting an end to subsidies – is a decision of inclusion and exclusion, a decision where something important is at stake. This opens up for the conflictual in the argumentation. Emma, who said before that her parents run a conventional pig farm and whose family's livelihood thus depends on agricultural subsidies, strongly disagrees with the farmer's standpoint and argues against it. The fact that, here, the student is clearly deeply personally engaged in the discussion (which is visible, for instance, in the raising of her voice) is an important difference with the above-described situation of political reflections where the political dimension emerges in a more 'distanced' way, i.e. as a systematic cognitive and rational inquiry into conflicting alternatives, without involving strong feelings and attachments.

Political moment

The last category in our typology, the *political moment*, is related to the 'moral reaction' in the ethical tendency (see chapter 6). Both situations are characterised by a strong feeling that affects you personally (a poignant experience). It is not planned in advance and not preceded by a rational consideration: it just happens to you, unexpectedly and out of the blue. Whereas a moral reaction concerns the moral dimension, a political moment has to do with strong feelings in relation to how to judge, prioritise, etc. in relation to questions of how to organise the society in the best way.

Notably, we found an example of a political moment in one of the student's answers on the questionnaire after the lesson that we characterised as a consensus-oriented deliberation (see 'Consensus-oriented deliberation'):

> At the same time the classmates' contributions sometimes make me worry about different worldviews and accordingly how different my own worldview other could run our world.

This reflection of the student reveals a strong feeling that struck the student suddenly during the deliberation that took place in the lesson. It demonstrates an emotional involvement in the issues that were addressed in the classroom discussion. Words such as 'worry' describe the gut feeling of an immediate experience of possible exclusion of one's own worldview in favour of others', very different, worldviews in decisions on how the world should be governed. The student's writings show how social life emerges here as constituting an 'us' and a 'them', the included and the excluded. Simultaneously, it becomes clear who are the opponents or the antagonists – in this case some of the classmates. It is the emergence of immediate, unpredictable and unexpected strong personal feelings that makes a political moment different from the closely related conflict-oriented deliberation. In a political moment, the participants are not only deeply engaged but the conflicting commitments and associated antagonistic relations are suddenly discovered and bodily felt (for a more detailed description, see Håkansson and Östman 2018)

Didactic implications

We summarise the different categories of the political tendency typology in Table 8.1.

We have emphasised the different character of each of these situations and the very diverse conditions for learning that they bring about. More specifically, our different examples above revealed large differences in the opportunities offered for students to develop political action competence when it comes to handling dissonant and conflicting voices in deliberation and decision-making regarding sustainability issues. If we want, as teachers, to be optimally prepared to adjust our teaching to specific purposes that we have in mind in view of developing political action competence, it is vital to be aware of this.

In educational situations that we labelled as *political norms*, qualification and socialisation (see chapter 4) are in the forefront. By initiating students in certain norms or rules that prescribe the democratically correct way of acting (e.g.

Table 8.1 Features that characterise the different situations of the political tendency typology

Political norms	• communicating rules that prescribe the democratically correct way of acting
	• exclusive focus on the form and process of participation in democracy
	• distribution of voices and power within the classroom and the school
Political reflections	• creating new or refined knowledge and understandings of the political dimension of sustainability issues
	• rational reflection
	• using prior knowledge and understandings
Normative deliberation	• taking a stand and defend it (argumentation)
	• orientation towards an a priori specified result as the only correct one
	• expected outcome is set by the teacher but not known in advance by the students
Consensus-oriented deliberation	• taking a stand and defending it (argumentation)
	• orientation towards coming to a consensus
	• resolving all divergence into one shared standpoint or decision
Conflict-oriented deliberation	• taking a stand and defending it (argumentation)
	• opening up for the conflictual
	• raising and defending opposing and contesting perspectives
Political moment	• strong bodily feeling (poignant experience)
	• not planned in advance and not preceded by a rational consideration
	• experience of antagonism regarding how to organise society under the condition of conflictual commitments and relations

(The rows "Normative deliberation", "Consensus-oriented deliberation", and "Conflict-oriented deliberation" are grouped under the vertical label **Political deliberation**.)

listening to everyone's opinions and ideas before making a final decision), we develop a particular kind of political action competence, i.e. students' ability to participate in democratic processes and to take on their role as a citizen according to these norms and rules. This is also a matter of socialisation, of transferring certain democratic values, attitudes, norms and worldviews, such as the belief that democracy is the preferred way of governance. Simultaneously, however, opportunities for person-formation are offered in the background, as a 'companion' (see chapter 4). Students might start to see themselves as an important change agent in the development of the school or as someone who wants to actively participate in politics (identification). Or, on the contrary, a student might – unintentionally – refuse to take part in the offered democratic participation setting, thereby dis-identifying him/herself from the roles implied in the political norms that are offered (subjectification).

With its focus on developing and refining knowledge and understandings through rational inquiry, *political reflections* are strongly oriented towards qualification. Students are encouraged to explore the political dimension of sustainability issues and learn about different and competing perspectives on how to organise society, about the decisions that are at stake, the ways in which public life can be governed, potential conflicts that might arise and modes of resolving them, power relations involved, etc. As explained above, in situations of political reflections the students will explore all this in a rather distanced, cognitive way. Thus, their learning will be rooted in intellectual problems rather than in a poignant experience such as a political moment. In chapter 10, we explain the difference between these diverse roots of learning in detail. Importantly, here too socialisation as well as person-formation can appear as companions through the knowledge content that is used for the foregrounded goal of qualification – socialisation, for instance, into the values of pluralism and respect for other opinions. Person-formation can occur when students are exploring a sustainability issue and are struck by its far-reaching impact to such an extent that it affects their identity-formation. They might realise, for example, that they are 'someone who cares for the planet'.

The categories of *political deliberation* show that experiencing the political dimension of sustainability issues in educational practice is not only a cognitive and rational matter but also involves highly valued commitments and standpoints.

In *normative* deliberation, the plurality of commitments and standpoints is downplayed by the orientation towards a specific outcome that the teacher has set in advance. A normative approach to ESE as the straightforward transfer of pre-set solutions as the only correct ones, however, has been criticised for violating the democratic and emancipatory purpose of education (see chapter 5). This risk is present both in situations of political norms and normative deliberation. If we as teachers instead encourage students to discuss norms and the motives behind them, we create opportunities to critically reflect on and influence those norms. After all, being qualified and socialised to act in accordance with certain norms does not necessarily mean that students are also personally affected or committed, nor that the acquired knowledge about how to behave will influence their behaviour in their life outside the confines of the educational context. This requires that concerns and commitments related to the content of education become part of the students' personality. Facilitating forms of deliberation where students are encouraged to think for themselves and to raise and defend personal standpoints can be a fruitful driver for such person-formation.

Through raising and defending personal standpoints – be it *consensus-oriented* or *conflict-oriented* deliberation – students can experience that questions regarding how to organise society in the face of sustainability problems involve decisions of which commitments to include and which to exclude and, hence, antagonism. They are offered the chance to discover and articulate their own and others' commitments, to be affected by others' argumentations, to learn to develop respect and openness for different perspectives in decision-making processes among competing commitments and to learn to understand that commitments are not always fully explainable or defendable by rational arguments only. This creates opportunities for person-formation: students might take on an identity connected to specific perspectives or commitments or, instead, realise that they are very different from the others (subjectification). Simultaneously, the students also learn things such as developing sound arguments (qualification) and attitudes such as listening to each other (socialisation).

As a teacher, it is crucial to realise that teaching about the political dimension of sustainability issues can always, unexpectedly, become a poignant bodily experience of antagonism and exclusion if a student experiences a *political moment*. As we will further explain in chapter 13, conflict-oriented deliberation can be a fruitful ground for political moments. Poignant experiences can be valuable and powerful sources for learning (see chapter 10). However, there are also risks involved in letting antagonism become manifest in the classroom, for example creating or strengthening conflictual social relations or, in the worst case, violence between students. It is therefore important that we consciously consider to what extent we want our students to become deeply personally engaged in discussions about disagreement, controversy, conflict, heavily invested commitments, etc. This requires context-sensitive consideration but also a certain competence in dealing with the political challenges involved in sustainable development teaching. Topics addressed elsewhere in this book can offer us guidance in this challenging task, for instance by elaborating how our interventions ('teacher moves') can direct students' activity either towards or away from conflict-oriented deliberation (chapter 13) or by explaining different possible ways to deal with 'political emotions' in the classroom (chapter 20).

Finally, it is important to highlight that, however well prepared we are, we can never predict or guarantee how the political dimension of sustainability issues will emerge in teaching and learning activities. As we saw in one of the examples discussed earlier, for instance, a consensus-oriented deliberation can always, unexpectedly, turn into a political moment. Or we might strive to involve our students in a conflict-oriented deliberation and find them persisting in striving to reach a consensus. Awareness of the variety of possible manifestations of the political tendency and the diverse opportunities for developing political action competence that these bring about can help us to flexibly adjust our teaching practices to changing or surprising situations.

References

Håkansson, M. and Östman, L. (2018) "The political dimension in ESE: The construction of a political moment model for analyzing bodily anchored political emotions in teaching and learning of the political dimension". *Environmental Education Research*, pre-published online: https://doi.org/10.1080/13504622.2017.1422113

Håkansson, M., Östman, L. and Van Poeck, K. (2018) "The political tendency in environmental and sustainability education". *European Educational Research Journal*, 17(1): 91–111.

Knutsson, B. (2013) "Swedish environmental and sustainability education research in the era of post-politics?" *Utbildning & Demokrati*, 22(2): 105–122.

Lundegård, I. and Wickman, P.-O. (2007) "Conflicts of interest: An indispensable element of education for sustainable development". *Environmental Education Research*, 13(1): 1–15.

Mouffe, C. (2013) *Agonistics: Thinking the World Politically*, Verso, London.

Öhman, J. and Östman, L. (2008) "Clarifying the ethical tendency in education for sustainable development practice: A Wittgenstein-inspired approach". *Canadian Journal of Environmental Education*, 13(1): 57–72.

Sund, L. and Öhman, J. (2014) "On the need to repoliticise environmental and sustainability education: Rethinking the postpolitical consensus". *Environmental Education Research*, 20(5): 639–659.

Van Poeck, K., Goeminne, G. and Vandenabeele, J. (2016) "Revisiting the democratic paradox of environmental and sustainability education: Sustainability issues as matters of concern". *Environmental Education Research*, 22(6): 1–21.

Wals, A. and Heymann, F. (2004) "Learning at the edge: Exploring the change potential of conflict in social learning for sustainable living". In Wenden, A. ed., *Educating for a Culture of Social and Ecological Peace*. SUNY Press, New York, 123–144.

9 Deliberation and agonism

Two different approaches to the political dimension of environmental and sustainability education

Ásgeir Tryggvason and Johan Öhman

Introduction: deliberation and agonism

To acknowledge the political dimension of environmental and sustainability education is to acknowledge that these issues cannot always be handled with more facts and information. Instead, the political dimension of sustainability issues points to how these issues are entangled with visions, hopes and opinions on how the society should be. In this sense, the political dimension does not only open up for normative questions, but also opens up a conflictual aspect where different perspectives on sustainability can clash into each other.

As seen in chapter 5, three different teaching traditions can be identified in environmental and sustainability education: the fact-based tradition, the normative tradition and the pluralistic tradition. Within the pluralistic tradition, the multitude of perspectives and conflicting opinions on sustainability issues are seen as the starting point, and not as an obstacle, for ESE practice. In this sense, the pluralistic tradition acknowledges the political dimension of environmental and sustainability education in terms of plurality, differences and conflicts. A crucial question is then: From what theoretical ground can teachers approach this plurality and these conflicts in their classrooms?

In this chapter, we provide a theoretical ground for approaching the political dimension of environmental and sustainability education. We do this by outlining *deliberation* and *agonism* as two different approaches to plurality and conflicts in classroom discussions. In the next section, we describe how pluralism and conflicts constitute a common ground for both the deliberative and the agonistic approaches. In the third section, we will describe some of the merits and disadvantages of a deliberative approach. Following this, the forth section outlines the merits and disadvantages of the agonistic approach. With these two ways for teachers to approach the political dimension, we discuss in the fifth and final section of this chapter why it is important to underscore their differences.

A common ground: pluralism and conflicts in ESE

The deliberative approach and the agonistic approach share a common ground in that they both acknowledge the pluralism of perspectives and conflicts as being key in teaching environmental and sustainability issues.

Pluralism of perspectives are by both the deliberative and the agonistic approaches seen as the starting point of democratic education. They both underscore that *politics* always involves different visions and opinions on how society should take form and contain different conceptions of what a good society means. Within ESE, this pluralism can be seen in how there exist plural ideas and visions of what a sustainable society should look like and how it should be achieved (Heymann and Wals 2002). In this way, the pluralism of political visions is at the heart of both the deliberative approach and the agonistic approach (Lundegård and Wickman 2012).

Conflicts between different visions and opinions on sustainability are within deliberation and agonism seen as the main starting point for authentic and fruitful classrooms discussions. Even if the two approaches have very different conceptions of what a conflict is and how it should be handled, they both clearly see conflicts as a place from which ESE can start (Englund et al. 2008; Sund and Öhman 2014). Moreover, within both approaches conflicts are not seen as just being an intellectual disagreement, but are instead understood as a phenomenon within political life that can involve a strong bodily experience where heated emotions always can erupt and intensify a conflict. A main difference between the deliberative approach and the agonistic approach can instead be found in their different conceptions of what role conflicts *should* play in education and classroom discussions (Tryggvason 2018). From a deliberative perspective, without conflicts there would be nothing to deliberate. What is important here is that the educational setting provides possibilities for different visions and opinions to be clearly formulated so that the differences between them can be outlined (Englund 2016). Moreover, the deliberative approach underscores that the differences between perspectives should not be superficial or based on misunderstandings or prejudices of what other perspectives imply. A similar idea is also emphasised by the agonistic approach when it places the differences between what people *want* instead of who they *are* in focus for democratic conflicts (Mouffe 2005). In this sense, the agonistic approach distinguishes between different kinds of conflicts, where some conflicts are compatible with democratic education while others are not (this will be explored in more detail later).

To sum up, the common ground for both the deliberative approach and the agonistic approach is that pluralism and conflicts constitute a starting point for classroom discussions. However, even with this common ground there are significant differences between deliberation and agonism, and in choosing between approaches, teachers need to consider both their merits and disadvantages.

Why deliberation?

Why should teachers take on a deliberative approach to environmental and sustainability issues in their teaching? What problems are there with the deliberative approach? In this section, we answer these questions by outlining the main characteristics of the deliberative approach as well as some points of criticism.

A *rational and respectful communication*

The deliberative approach draws on a communicative ideal in which rational arguments and respect for each other are emphasised. The approach mainly stems from the German philosopher Jürgen Habermas and his influential work on communicative rationality and deliberative democracy. A core component of the deliberative approach is the idea of rational arguments. Stressing the role of rational arguments in classroom discussions means that the participants should aim to avoid heated emotions and personal conflicts. Conflicts in the classroom should instead, according to the deliberative approach, stem from the differences between opinion and arguments. The deliberative theorist John S. Dryzek (2005) underscores that it is difficult to change one's mind 'if one's position is tied to one's identity' (p. 229). In line with this, Tomas Englund (2016) argues that it is important to put the issue itself in the foreground, rather than students' different identities. In that way, it is the issue and the argument, and not the person, which is in the centre of the deliberative process. Moreover, by focusing on the issue and the arguments, rather than on the person, arguments based on tradition or a person's status lose some of their grip. Thus, an argument is not valid just because the tradition says so, or because it is a person with status that has formulated the argument.

In this way, the deliberative approach outlines a communicative setting in the classroom that places charismatic seduction and populist rhetoric in the background. This ideal, which stems from Habermas's theory of communicative rationality, places faith in the force of the better argument itself, rather than relying on an authority or on a well-spoken agitator (see Habermas 1996; see also Lundegård and Wickman 2012, 154–155).

Even though the deliberative ideal emphasises a rational communication, centred on the argument and not the person, it requires a solid respect between the participants deliberating. The deliberative approach claims that students' emotions and cultural belonging should not be given a main role in the deliberative process; in other words, it is not 'you' but 'your arguments' that matter. Nevertheless, there are no arguments without persons formulating them, and there can be no deliberation without participants. This implies that it is crucial that the participants respect each other even when they strongly disagree on a topic. According to Englund (2016), a key component of the deliberative approach is the respect for the *concrete other*. Here, the emphasis on the concreteness of the other is important. This means that the respect for the other should not depend on his or her social status or on his or her status as a representative for a political position. Instead, the respect is about the other as a concrete person of 'flesh and bone' with whom one deliberates. This prompts for an equality between participants, as the respect is independent of the person's social and political position. For example, if an argument in a classroom discussion is put forward by the teacher, by a popular student or by someone who rarely says something, the respect for the person should be there regardless of these social aspects. What is scrutinised, criticised and discussed is the argument itself, and not the person formulating it. This approach could be seen as being suitable for classroom dis-cussions as it aims to enable students to openly make up their mind based on the arguments put forward in the on-going discussion.

A critical point to these ideas of rationality and respect is that students' cannot leave their identity at the door when they enter the classroom. Politics, a critic could say, is bound up with who we are and what we want, and identities and belongings are vital parts of democracy and citizenship. As political discussion concerns what 'we' want, it is not possible to leave the collective identity that constitutes this 'we' out of it (Ljunggren 2010; Tryggvason 2018; Zembylas 2011).

Aiming for consensus in the classroom

A main characteristic in the deliberative approach is the idea that participants in a discussion should aim toward a consensus. However, aiming toward a con-sensus in discussions over sustainability issues should not be understood as an aim to reach a final conclusion or reach a conclusion that will end all discussions (Englund et al. 2008). The idea is instead that discussions over burning political issues are dependent on decisions, therefore discussion cannot go on forever. To reach a decision that all participants can agree upon is to establish a consen-sus. Such a consensus does not have to be everlasting but is always a temporary decision that can be reconsidered through a deliberative process. One could say that this idea of reaching a temporary decision becomes particularly important in environmental and sustainability issues, as there is an apparent lack of time in many of the issues, as well as an urgency to go from discussions to actions.

From the perspective of teaching, the idea of creating communicative situ-ations where students' aim to agree with each other on difficult and complex issues can be seen as a suitable educational goal. To encourage students to find common ground between conflicting views and enable them to form a collective will is a key competence in democratic societies. From a deliberative perspec-tive, it is therefore reasonable that students are given opportunities to develop such competencies (Gutmann 1987). The aim to reach for consensus in class-room discussions is from a deliberative stance an *educational* qualification of the discussion. Conflicts between different visions of society is a starting point for discussions, but as Englund (2016) argues, education should not settle with just recognising these differences but aim to qualify the communication by aiming for a collective will-formation. Moreover, the deliberative approach does not claim that it is always possible to reach a consensus; at the end of the day it is always an open question whether a consensus is reached. But the idea is that it is desir-able to *aim* for consensus between conflicting views in classroom discussions over environmental and sustainability issues. This goes back to the idea of rationality, as we described above. To be able to reach a consensus in a classroom discussion is dependent on whether the students are able to change their minds. If no one is able to change their mind, a consensus is hard to reach, and to change one's mind is easier if the opinion is not tightly connected to one's identity or cultural belonging. Thus, engaging in a discussion from a deliberative perspective should therefore not be about putting your 'self' at stake, but only your opinions and argument.

A critical perspective on the idea of aiming for consensus is that it tends to subvert conflicts that otherwise could be brought up for discussion. From such a critical perspective, consensus tends to close down communication and the fruitful interplay between conflicting views (McKenzie et al. 2015). An example of this has been found in empirical studies of classroom discussions of sustainability issues. Michael Håkansson et al. (2018) studied how students in an upper secondary school discussed environmental and sustainability issues (see chapter 8). When the discussion was held in the classroom involving all the students in the class, a strong consensus prevailed. Thus, the students all seemed to agree with each other, even on issues that are highly controversial in the public debate outside school. From a deliberative perspective, this could indicate that a rational consensus was established in the discussion. However, when the students were asked individually (and anonymously) about the classroom discussion, heated conflict between ideas and perspective became evident. Behind the prevailing consensus, conflicts were lurking both between students and between different perspectives on sustainability. From a critical perspective, this could be understood as an example of how a consensus can suppress conflicting views on burning issues. In that sense, the aim for consensus can become an obstacle for different perspectives to arise. In a broader perspective, this is also a crucial problem that the deliberative approach has to handle, namely, How can the pluralism of ideas that constitutes the starting point for the deliberative approach be maintained alongside the aim to reach consensus?

Why agonism?

The agonistic approach is an approach that is partly formulated as a response to the deliberative approach and much of its theoretical development has taken shape in relation to the deliberative approach. In our outline of the agonistic approach, we will therefore refer back to the deliberative approach in order to show the contours of agonism.

Emotions as part of the political dimension

A starting point for the agonistic approach is that emotions and identities are vital parts of a vibrant democratic and political life. To be engaged and involved in environmental and sustainability issues is to be emotionally engaged in these issues, in other words to be engaged is to feel that these issues matter. It is hard, for example, to imagine a student who is highly engaged in a sustainability issue without being emotionally invested. Without the emotional investment, an engagement in an issue would not appear authentic. From an agonistic perspective, it is *because* emotions are an integrated part of political and democratic life that they also need to be seen as an integrated and important part of democratic education (Ruitenberg 2009). When teachers approach the political dimension of ESE, they will inevitably face the question of *how* to handle students' emotions in the classroom concerning environmental and sustainability issues (Sund and Öhman 2014; chapter 8 and 20).

The agonistic approach to emotions in the ESE classroom takes it starting point in the distinction between moral emotions and political emotions. The difference between a moral emotion and a political emotion is that they are directed toward different objects (Ruitenberg 2009). For example, being upset over how one's siblings treat the family cat could be seen as a moral emotion, as it is directed towards a personal object and a personal relation. In contrast, being upset over the loss of biological diversity that follows from the deforesting of the Amazon is a political emotion, as it is directed toward a societal object (Ruitenberg 2009). In approaching the political dimension of ESE practice, it is primarily the latter kind of emotions (political emotions) that are relevant to frame as productive elements in the classroom.

Against the background of this distinction between moral and political emotions, the agonistic approach claims that it is crucial that political emotions that arise in classrooms are given a legitimate place and are not suppressed or downplayed. These emotions hold a great amount of potential in teaching about environmental and sustainability issues, as they are key in getting students engaged in issues that will have bearing on their present and future society.

A critique of the agonistic approach is that emotions in classroom discussions carry the risk of turning the discussion into conflicts between students (Englund 2016). For instance, a heated discussion of the loss of biological diversity, and where the responsibility for this lies, could turn into an emotionally charged conflict that revolves around the students themselves rather than the issue itself. Moreover, if emotions can turn political conflicts into personal conflicts between students, it is reasonable to think that it will be more difficult for students to change their minds on an issue if they are emotionally invested in it (Samuelsson 2016). To establish a communicative setting where students jointly can explore an issue and its different solutions will be difficult if emotions are endorsed to run high. From this critical perspective, the downplaying of emotions is not primarily about excluding something from discussions but is instead about making room for other things, such as students' joint exploration, problem solving, creativity and curiosity. These things, the critic could say, are crucial components of education, and it would therefore be problematic if they were excluded in classroom discussions when emotions are endorsed to run high.

Aiming for democratic conflicts in the classroom

The agonistic approach does not aim for consensus, instead it aims at enabling conflicts to be played out politically and democratically, rather than in a moral and a rational register (cf. Todd 2010). A main question is, therefore, What distinguishes democratic conflicts from undemocratic conflicts? The answer to this lies at the very core of the agonistic approach, namely democratic conflict is where the 'other' is seen as an *agonistic adversary*, while undemocratic conflict is where the 'other' is seen as an *antagonistic enemy* (Mouffe 2005). Being each other's adversary means that the conflicts between 'us' and 'them' stems from difference between what 'we' want in contrast to what 'they' want. Such a conflict does not revolve around ideas about who the others *are* essentially, but are about what they *want* (Zembylas 2011; chapter 20). The agonistic approach therefore aims to enable conflicts to have a democratic outlet where the difference between 'us' and 'them' are about different visions of society (Todd 2010). If such democratic outlets are not open, there is a risk that conflicts will become antagonistic and undemocratic, where the other is seen as an enemy (Mouffe 2005).

Bringing this idea of conflicts into ESE practice can be illustrated if we return to the empirical study mentioned above about consensus and conflicts (Håkansson et al. 2018; chapter 8). In that classroom, conflicts were present but not handled or discussed openly; instead, students kept their emotions and disagreement to themselves. From an agonistic perspective, one could say that the conflicts were not given a democratic outlet or the possibility to be formulated within the overarching consensus that prevailed in the classroom. If students are not given the possibility to frame and formulate their disagreements with each other as a democratic conflict, then there is an overarching risk that the disagreements will turn into undemocratic antagonistic conflicts. The possibilities to formulate a disagreement as a democratic conflict between adversaries is not something that will arise automatically, but requires practice, concepts and competencies; in other words, it requires teaching. Therefore, from the agonistic perspective, it is crucial that education aims to provide students with the possibility of positioning themselves as each other's adversaries (Ruitenberg 2009).

Let us now turn to a few critical points that the agonistic approach faces. A main critique that can be directed toward the agonistic approach is that education is not the same as politics. Put differently, even if an agonistic approach could be seen as a suitable model for conflicts and disagreements within political organisations or in a parliament, this does not entail that it is a suitable model for classroom discussions. In their book *The Political Classroom*, Diana Hess and Paula McAvoy (2015) argue that a classroom is an 'unusual political space' and it cannot be easily compared to other political spaces. Thus, theories of politics cannot be implemented undistorted as educational theories. What makes classrooms into unusual political spaces is that students are not always there voluntarily, often they are obligated to participate and they are often encouraged to discuss issues that they have not chosen themselves. Furthermore, they are obliged to do this with peers they will spend the rest of their school year with, if not more (Hess and McAvoy 2015). This could sound a bit gloomy, but our point here is to highlight that classroom discussions differ from other gatherings that discuss environmental and sustainability issues (as well as other political issues). This means that whatever *democratic* advantages that the agonistic approach has to offer, they need to be evaluated against their *educational* consequences. Students do not only enter the classroom and the discussion, they also leave the classroom; they go to recess, to other classes and to lunch together. As the classroom is this sense is an 'unusual political space', it requires unusual care and consideration of how emotions, identities and conflicts are handled. A critic, taking a deliberative stance, could say that it is exactly this uniqueness of the classroom that makes it a suitable place for students to explore issues from different perspectives and together weigh the arguments against one and other (Englund 2016). The classroom is also an unusual political space in a more concrete sense. As a communicative setting, the classroom is one of the few places in society where young people have the opportunity to reach a temporary consensus in burning political issues. The opportunities of maintaining conflicts, which the agonistic approach promotes, seems instead to be endless (cf. Wildemeersch and Vandenabeele 2005).

Discussion: why not both?

The political dimension of teaching environmental and sustainability issues is a dimension where diverging hopes, dreams and fears of the common future both encounter and clash into each other. In the midst of this complexity stands the teacher. Choosing an approach, whether it is deliberation or agonism, falls back on the question of what kind of experience the teacher wants to promote. Should the lesson be about the joint inquiry into different perspectives, where the students experience the formation of a rational consensus? Or should the lesson instead be about enabling the students to position themselves as adversaries, where they experience the transformation of disagreements into democratic conflicts? In answering these questions, the purpose of the lesson, as well as the teacher's contextual awareness and knowledge about the students, are of course crucial (see chapter 11). What we want to underscore here is therefore not which approach to prefer, but instead draw attention to the differences between the two approaches.

By characterising deliberation and agonism as two *different* approaches, the teacher's alternatives becomes theoretically clear, even if both the decisions and alternatives in practice can be much cloudier. In the research field of democratic education, there has been an on-going debate about whether deliberation and agonism are incompatible (Englund 2016; Tryggvason 2018). A main thrust in this debate is how to understand the nature of conflicts in both society and classrooms. In short, from a deliberative perspective conflicts are seen as a vital part of democracy, but at the same time as a phenomenon that can be overcome. Likewise, from the agonistic perspective, conflicts are seen as vital to democracy, but also as constitutive part of human societies that cannot be totally eroded or overcome through consensus (Mouffe 2005). Putting this debate to the side for an educational perspective, one could say that there is not only a difference in how the two approaches understand conflicts, but also a clear difference in how they understand *the classroom*. From the agonistic standpoint, one can criticise the deliberative approach for believing that students can leave their emotions and identities at the door when they enter the classroom. Similarly, a proponent of the deliberative approach can point to how students not only enter a classroom but also leave it, meaning that the agonistic approach does not acknowledge that students must be able to get along with each other as they leave the classroom and go to other educational activities together. From this deliberative standpoint, being each other's adversary cannot be seen as a desirable relation for students' shared life in school.

To conclude, in this chapter we provide a theoretical ground for teachers to approach the political dimension in environmental and sustainability education. Deliberation and agonism can be seen as two approaches that give the teacher different didactical tools in staging education. How this 'stage' will take shape, and what kind of communication that it will enable, depends on how the teacher approaches the plurality and conflicts of environmental and sustainability issues.

References

Dryzek, J.S. (2005) "Deliberative democracy in divided societies. Alternatives to agonism and analgesia". *Political Theory*, 33(2): 218–242.

Englund, T. (2016) "On moral education through deliberative communication". *Journal of Curriculum Studies*, 48(1): 58–76.

Englund, T., Öhman, J. and Östman, L. (2008) "Deliberative communication for sustainability? A Habermas-inspired pluralistic approach". In Gough, S. and Stables, A. eds., *Sustainability and Security Within Liberal Societies. Learning to Live With the Future*. Routledge, London, 29–48.

Gutmann, A. (1987) *Democratic Education*. Princeton University Press, Princeton.

Habermas, J. (1996) *Between Facts and Norms. Contributions to a Discourse Theory of Law and Democracy*. MIT Press, Cambridge, MA.

Håkansson, M., Östman, L. and Van Poeck, K. (2018) "The political tendency in environmental and sustainability education". *European Educational Research Journal*, 17(1): 91–111.

Hess, D. and McAvoy, P. (2015) *The Political Classroom. Evidence and Ethics in Democratic Education*. Routledge, New York.

Heymann, F. and Wals, A. (2002) "Cultivating conflict and pluralism through dialogical deconstruction". In Leeuwis, C., Pyburn, R. and Röling, N.G. eds., *Wheelbarrows Full of Frogs: Social Learning in Rural Resource Management: International research and Reflections*. Van Gorcum, Assen, 233–241.

Ljunggren, C. (2010) "Agonistic recognition in education: On Arendt's qualification of political and moral meaning". *Studies in Philosophy and Education*, 29(1): 19–33.

Lundegård, I. and Wickman, P.-O. (2012) "It takes two to tango: Studying how students constitute political subjects in discourses on sustainable development". *Environmental Education Research*, 18(2): 153–169.

McKenzie, M., Bieler, A. and McNeil, R. (2015) "Education policy mobility: Reimagining sustainability in neoliberal times". *Environmental Education Research*, 21(3): 319–337.

Mouffe, C. (2005) *On the Political*. Routledge, New York.

Ruitenberg, C.W. (2009) "Educating political adversaries: Chantal Mouffe and radical democratic citizenship education". *Studies in Philosophy and Education*, 28(3): 269–281.

Samuelsson, M. (2016) "Education for deliberative democracy: A typology of classroom discussions". *Democracy & Education*, 24(1): 1–9.

Sund, L. and Öhman, J. (2014) "On the need to repoliticise environmental and sustainability education: Rethinking the postpolitical consensus". *Environmental Education Research*, 20(5): 639–659.

Todd, S. (2010) "Living in a dissonant world: Toward an agonistic cosmopolitics for education". *Studies in Philosophy and Education*, 29(2): 213–228.

Tryggvason, Á. (2018) "Democratic education and agonism: Exploring the critique from deliberative theory". *Democracy & Education*, 26(1): 1–9.

Wildemeersch, D. and Vandenabeele, J. (2005) "Relocating social learning as a democratic practice". In van der Veen, R., Wildemeersch, D. Youngblood, J. and Marsick, V. eds., *Democratic Practices as Learning Opportunities*. Sense Publishers, Rotterdam, 19–32.

Zembylas, M. (2011) "Ethnic division in Cyprus and a policy initiative on promoting peaceful coexistence: Toward an agonistic democracy for citizenship education". *Education, Citizenship and Social Justice*, 6(1): 53–67.

Part III

Designing and implementing teaching and learning practices

10 A transactional theory on sustainability learning

Leif Östman, Katrien Van Poeck and Johan Öhman

Introduction

In this chapter, we further develop the transactional theory on learning that is introduced in chapter 3. We address the specific content that we should pay attention to as teachers in an environmental and sustainability education (ESE) context: learning sustainability-related habits that allow for creativity. We also present models for understanding learning processes in an ESE context: two different 'routes' learning can take (i.e. short and long learning loops) as well as three different 'roots' for learning (i.e. intellectual disruption, changes in the physical surrounding and poignant experiences). Furthermore, we elaborate on four crucial aspects that influence the learning outcomes: the intrapersonal, the interpersonal, the institutional and the physical. The presented models and aspects can be used as a background for designing efficient and fruitful ESE teaching, which will be further developed in chapter 11. The chapter ends with a reflection on the relation between bodily feelings and cognition, since an important part of ESE learning concerns not only knowledge but also feelings connected to ethical and political issues.

Learning a habit

Most of the time we live and act in accordance to acquired habits, which means that we mainly act in everyday life without reflecting. If one has acquired the habit to shop sustainably, for instance, this means that one will habitually buy certain products and not others without consciously reflecting on these purchases and the criteria to select them time after time. A habit can generally be described as a predisposition to act in a certain way in specific activities. More concretely, a habit contains a specific way of coordinating with the surrounding world in relation to the purpose that governs the activity. In the example of sustainable consumption, this means that if one does the weekly shopping, one might go to a certain supermarket that offers a wide range of sustainable products and habitually fill the shopping cart with eco-certified organic food, fair trade coffee, FSC-labelled paper, etc. One knows where to find these products in the supermarket and only needs to look superficially at the packaging and sustainability labels to take them from the racks.

Every habit consists of two elements: a specific attentiveness and coordination with the environment. A habit is focused on *certain* objects (e.g. eco-labels), which means that a *specific* attention is involved. This is important for our teaching. If we want our students to become competent in, for example, political reflection and teach them how to analyse and discuss different standpoints on sustainability issues (see further chapter 8), they need to develop a habitual attention, i.e. pay attention to certain objects. If they read and discuss a text on immigration, for example, it is the different standpoints and the arguments that they should pay attention to, but not the spelling of the words. Thus, it is important for the students to learn to habitually select certain objects from the surrounding world (physical – e.g. a book – and social – e.g. classmates' utterances) to employ in the activity. All the other objects should be neglected. John Dewey (1938/1997) calls this 'the *environment*': the material for doing things with in order to achieve something, and he distinguishes this from the totality of objects at reach in the activity: the *surrounding*. Students are supposed to learn to habitually coordinate their actions with the created environment in such a way that certain expectations (purposes) are fulfilled. To learn a habit is to learn to, habitually, stage a relevant environment and to intellectually reason and bodily act in relation to that environment in such a way that certain outcomes are created.

In order to illustrate the learning of habits, we will look at a concrete classroom example. It concerns an activity where upper secondary students are supposed to argue for certain solutions regarding climate change (Rudsberg and Öhman 2015, see also chapter 14). In the following utterance, Kate is responding to a student who claimed that oil-producing countries need to convert to other sources of income. Dubai is given as an example as the city invested in tourism as an alternative income.

Kate: That's not so environmentally friendly either. In the first place you have to fly there and then you can't live in a city in a sustainable way. [Inaudible] but in the desert it's even harder, because then you have to pump groundwater all the time.

Taking into account that Kate was supposed to argue for solutions for climate change, her selection of objects – transportation (flying), need of water, specific values (environmentally friendly, sustainable) – is highly relevant in relation to tourism and climate change. She has learned a suitable attentiveness and thus creates a relevant environment for the task. She also does something with this environment. She coordinates the objects in such a way that they become part of an argumentative structure – she formulates a counterargument to the first student's claim – in accordance to the expectation of the activity.

To summarise what has been said so far, the content to be learned when learning a habit is the following:

a   The intellectual and bodily ability to stage a relevant environment out of the surrounding world, i.e. to master the selective *attentiveness* necessary for putting relevant objects in the world in focus for the activity.

b   The intellectual and bodily ability to *coordinate* the staged environment in relation to a purpose in order to achieve an expected outcome. Such ability includes relevant knowledge, skills, judgements, reasoning, social communication, bodily movements, etc.

A very important aspect of learning a habit – especially in the context of sustainability issues – is judgement. As it is impossible to equally involve all available knowledge in a decision or in an activity, we need to make a hierarchisation (prioritisation) amongst different knowledge emanating from different perspectives (social, ecological and economic). In the case above, Kate could also have involved economic and cultural arguments, but she settled with mainly ecological ones. Thus she made a certain prioritisation. When doing so, one always needs to involve values in one form or another (see further chapter 3). The values involved can be practical or aesthetical but also moral or political. For Kate, the environmental-moral value was in focus. To learn a selective attention thus also implies learning to make judgements. Kate focused on certain objects, while many others would have been possible too, e.g. inequality, immigrants' working conditions and cultural exchange through tourism.

Since the learning of a specific attentiveness and a specific coordination with the environment involves judgements, one can say that learning habits concerns both qualification (knowledge and skills) and socialisation (values, judgements). Below, we will demonstrate that it also includes person-formation (see chapter 4).

Habits and creativity

One of the major tasks of ESE is that students learn habits that allow for creativity. Creativity is important when we encounter sustainability problems that are new for us, because we then need to use imagination together with knowledge to come up with innovations. In the following, we elaborate the specific content that students need to learn in order to develop habits that allow for creativity.

What is important to recognise is that a habit can have different qualities. Some can be called plastic; others are frozen. The latter can be called routines. Routines are rigid because the coordination with the environment is static and does not allow adjustments to changes in the environment (Dewey 1916/1997). As we become unaware of them and do not control them, they control us. In a routine, we thus have small possibilities to handle surprises and here the possibility for learning and thereby developing the habit is limited in scope. If the habit is plastic, however, the possibilities for adjusting our actions in relation to changing surroundings and self-selected aims are bigger. Therefore, the possibility for learning is also higher than in the first case (see also chapter 3).

A plastic habit allows the students to be creative. If the students are supposed to learn plastic habits instead of frozen routines, it is not enough that they learn the ability to coordinate the staged environment in relation to the purpose of the activity, for example learning how to sort out recyclable waste at school. They also need to be able to involve their artistic imagination when coordinating with the environment. This implies involving their personal creative visions when, for example, designing solutions for how to take care of waste at school (Andersson et al. 2018). An example of creativity was observed in a small school in western Mongolia (see Östman et al. 2013). The community did not have any waste management or recycling programme. The school worked with the model of 'locally relevant teaching' (LORET – see chapter 3). Teachers and students decided together to create an innovation for sustainable local waste management that the community could later take over. The students studied the issue of waste management in different subjects and came up with the following innovations: to create waste bins, to distribute them in the community and to create a programme for collecting and recycling. Part of the programme was that the school's cleaning personnel emptied the waste bins and took the garbage to the school, where the students sorted it. Plastic bottles and glass were sent to the community centre for recycling and the rest was picked up by a lorry to be transported to the waste dump. The students' imagination and its actualisation created a whole system and programme for waste management, which no one had done before in the community. This original and creative innovation became possible because participants involved their artistic imagination and combined it with knowledge learned in different subjects. What is important to recognise is that with artistic imagination, we do not mean fantasies that are not connected to the problem and the actuality, but rather we mean to envision the possible starting with the actual situation.

If we return to the content of learning a habit, point a and b above should be complemented if we want students to develop creativity:

c   The intellectual and bodily ability to use imagination and knowledge in coordinating the staged environment in order to achieve a personal artistic vision.

An artistic expression is personal: it is a result of a personal engagement (see further Andersson et al. 2018). A consequence of students expressing themselves artistically is that they step out of the role of students and instead show their uniqueness as a person. Thus when learning plastic habits, the students also develop themselves as persons. The Mongolian students did not only develop solutions for waste management, but also commitment to the recycling programme and the waste bins they designed and probably pride in being able to solve a sustainability problem for the community. They did not create the waste bins only to pass an exam or to get good grades. They were authentically engaged and, as such, put themselves at stake. If it would not be successful, they would be affected personally and emotionally.

As mentioned above, the learning of habits does not concern only *qualification* (knowledge and skills) and *socialisation* (values, judgements). What we illustrated with the Mongolian example is that also *person-formation* is part of that learning (see chapter 4).

Learning in short and long loops

We start to reflect and think when we encounter problems, i.e. situations where we can no longer use our earlier acquired knowledge, skills and values to proceed as usual in an activity (see chapter 3). Without problems, we live our lives according to habituated beliefs. But sometimes we encounter a situation where our habits become disturbed. We stop the activity in which we are engaged and start to reflect. The disturbance of habits wakes us up from non-reflectivity (Figure 10.1). We then experience a 'problematic situation' (Dewey 1938/1986). This becomes noticeable, as we can no longer act immediately: we hesitate, sigh, etc. There are two 'exits' of such a problematic situation, which we call the short and the long learning loop. A *short learning loop* occurs when students can easily solve the situation with the help of resources belonging to their habit. In our example of habitually buying sustainable products, this could happen when, suddenly, one of the usual eco-certified products is no longer available. A short learning loop could then result in scanning the racks searching and finding an alternative product with another eco-label. Short learning loops can result in learning in terms of consolidating and enriching the habit (see Figure 10.1). But sometimes the problem is harder to tackle and requires considerable time and energy. Then, we enter a *long learning loop* that involves an 'inquiry' (Dewey 1938/1986). If the inquiry is successful, the learning outcome is more extensive than in the short loop, involving a substantial transformation or even the start of a new habit. Through experimentation we try to solve the problem, and the result is, if we are successful, new knowledge, skills and values. Thus, these new reflective outcomes re-create (transform) a habit or create a new habit (see Figure 10.1).

The whole process of tackling problematic situations requires a sense for what the problem is and how it can be solved in this situation and in this activity, but without being able to explain why. Sometimes we call such a sense intuition. In school, we can often see whether the students have a sense of the problematic situation: if not,

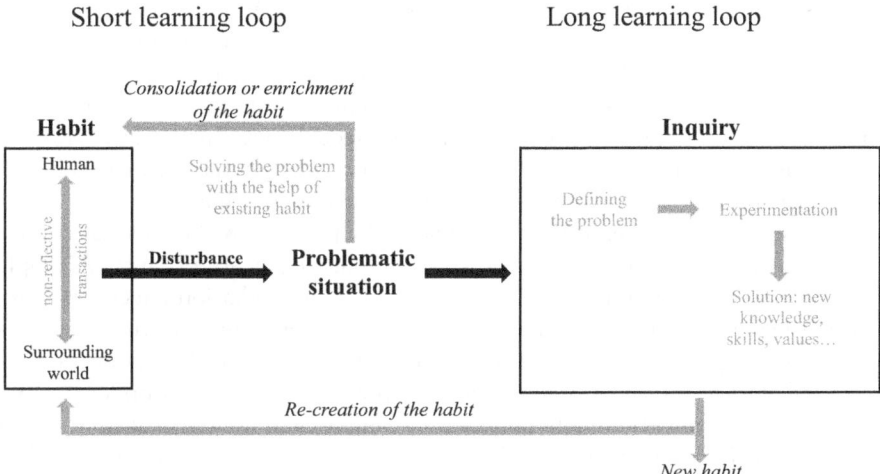

Figure 10.1 The model shows the process from disturbance of a habit, the establishment of a problematic situation and the continuation through a short and long learning loop respectively

Source: Author's own.

they just throw out guesses, one wilder than the other. When sense is present, genuine experimentation can occur. Without it, it becomes just a guessing game.

To define a problem through inquiry is not as easy as it might seem in Figure 10.1. The definition of the problem often starts with a sense of the problem, followed by experimentation. During the experimentation, the problem becomes clearer step by step. Thus, experimentation often goes hand in hand with the process of clarifying the problem. In experimentation, students need to spend more time and energy compared with the short learning loop, because here the students will start reflections and discussions, come up with ideas, plans, etc. They might need new information, they might need to ask someone for advice, and they need to apply the result of their work in order to solve the problematic situation.

Different types of disturbances create a learning situation

Above we highlighted that disturbances are an important start for learning. We will here elaborate on three different types of disturbances, three different 'roots' of learning, so to speak:

1   An intellectual disruption.
2   A change in the physical surroundings.
3   A poignant experience.

The first type of disturbance is a disruption in our understanding, which means that we cannot proceed with the activity and instead start to reflect. This could be, for instance, someone telling us that s/he saw a documentary on a specific eco-label where it was argued that the criteria to obtain it are not as strict as they should be to guarantee genuinely sustainable products. This could cause a disturbance of our habit to buy eco-certified products, as we are no longer sure that just checking for eco-labels is a sufficient strategy for sustainable consumption. In order to (re)gain assurance and understanding of sustainable shopping, we need to start an inquiry. Is the information correct? What exactly is the problem addressed in the documentary? Are there other eco-labels that are more reliable? This may result in a change or refinement of our shopping habit.

The second root of learning is a change in the physical world, and that change is affecting our body, which creates a feeling that we pay attention to. A simple example: When the sun comes out of the clouds, the radiation affects our skin, which makes the atoms in the skin move faster. Sometimes we recognise this affect (feeling hot as we transpire) and we might become puzzled about the feeling and start to reflect on it: 'the sun is so hot nowadays . . . I wonder if it has something to do with climate change'. We might even draw conclusions of our reflections: 'hmm, next time I will bring a long-sleeved shirt to protect myself' or 'I need to check if it has something to do with climate change'. Thus, the bodily feeling created a puzzlement (a problematic situation) that can lead to learning. What is crucial to recognise is that this example also illustrates that learning is not merely the product of human intentions: changes in the physical surroundings can also trigger learning.

The third type of disturbance is a poignant experience. A poignant feeling is very strong, is often existential and can be described as, for example, worry, anger, happiness, sadness, etc. (see below). Furthermore, a poignant experience is unanticipated and unexpected; it comes out of the blue. The example that follows illustrates a poignant experience in terms of a moral reaction (see chapter 6; example collected from Öhman and Östman 2007). Two women, taking part in an ecology class at a university, are having the following conversation while collecting animals on the shore in order to later put them in an aquarium for scientific studies:

Karin: Hell, for crying out loud. It feels awful when you pull them loose.
Ellen: So what is it?
Karin: A sea urchin.
Ellen: It's stuck.
Karin: I don't know. It seems weird. We've got to learn to pick them off with our hands.

Karin is having a moral reaction, i.e. she is suddenly and unexpectedly experiencing a strong bodily feeling of dislike, as her first comment reveals. It is obvious that it is a poignant experience that concerns how to treat the sea urchin. The evaluation of dislike breaks out in the scientific activity of collecting animals, and she explains to the teacher that she does not want to continue with collecting animals as a preparation of the scientific study. The problematic situation that was created in this situation can be called a moral dilemma, because the course requires Karen and the others to collect animals as part of the investigation.

Since we are already from birth (and probably even before that) affected by the physical and cultural world, we experience feelings very often. What is important to notice is that a bodily feeling is unfixed, vague, inexact, unstructured and something that we cannot pinpoint or define with language (Gould 2010, 26–27). In pragmatist as well as in phenomenological terminology, another term for feeling is immediate experience. An immediate experience:

> is our experience as it is 'given', that is, as it is before we start tampering with it, interpreting or editing it, actively working on it in some sense.
> (Overgaard 2008, 291)

It is first when we start to actively work on the immediate experience, the feeling that we can start to describe it and make it intelligible. In the case of Karen, the first experience was a strong fuzzy feeling – 'it feels awful' – continued by a reflection – 'I don't know. It seems weird. We've got to learn to pick them off with our hands'. The reflection was followed by a decision, i.e. that she did not want to continue to collect animals.

In sum, a problematic situation can occur through a disruption of intelligibility, a change in the physical surroundings or a poignant experience. Each of these makes the students stop and reflect. What is important to mention here is that a poignant experience and also sometimes a less strong bodily feeling disturbs the initial plan of the teacher, leading to unexpected learning. Below we continue by elaborating in more detail the learning that occurs in short and long learning loops after a disturbance has happened.

Learning as bridging gaps and creating relations

When students experience a disturbance of their habits (see Figure 10.1), a gap occurs between their earlier experiences (knowledge, skills and values) and the new situation, i.e. a new object (for example the word antagonism) or a feeling (for example uneasiness). Thus students cannot make the new object or the bodily feeling intelligible with the help of their earlier acquired knowledge, skills, values, etc. Formulated a bit differently: The students cannot create relations between their old experience and the encountered situation. Without being able to create such relations, they cannot continue with the activity they are involved in. And hence they cannot achieve the expected learning outcome of the activity.

It is important to differentiate between different types of gaps. Students experience gaps all the time because they are constantly encountering new objects or feelings. Many of the gaps are immediately bridged, but sometimes a gap makes them stop the activity and start to reflect. In such a case a problematic situation has occurred. If students can, with the help of their old habit, solve the problematic situation, then they are back on track quite easily. This is what happens in the short learning loop. But if they do not succeed, they need to start up an inquiry. An inquiry is thus needed if the gap is lingering, and here they enter the long learning loop.

In order to bridge gaps relations must be created. Relations between what stand fast for the students – what they already know, value, etc. – and the new situation. The new situation can be a bodily feeling (for example a poignant experience) or a new object (for example an eco-label). By creating relations to what stands fast for them, the students are creating intellectual and bodily connections to the world. Let us return to the example with the sea urchin.

It is obvious that there are many experiences that stand fast for Karin, i.e. there is a lot of knowledge, skills and emotions that she has learned before this conversation happens. For example, she knows how it feels when something is 'awful', she knows what a sea urchin is, she knows what to do when picking animals out of the water and she knows what it means when something is 'stuck' or feels 'weird'. And she knows what the purpose of the activity is (to collect animals with their hands). She is encountering the sea urchin when she is trying to pick it up with her hands. In this encounter, she gets a strong existential bodily feeling: 'Hell, for crying out loud'. She then creates a relation between the bodily feeling, awful and pull them loose. All parts in this relation (bodily feeling – awful – pull them loose) stand fast for Karin, but the relation is new and, furthermore, totally unexpected. She manages to make the bodily reaction meaningful by creating a relation between the action to pull them loose and her old experiences of the feeling of awfulness. What is important to notice is that in the case of Karin, the disturbance created a poignant experience that broke with her old habit: she was a student of ecology at university level and she therefore had a long experience of scientific investigations where manipulating nature is a natural part of such investigations. Because of this break, she had to take a long loop in order to make the experience intelligible. After the conversation with Ellen, she went to talk to the teacher in order to continue to make the experience meaningful in terms of re-creating the old habit: she wanted to continue to study ecology but not to collect animals. Interestingly, Karin discovered a new self because of this poignant experience. She no longer wants to take on the role of a manipulator of animals that is offered through the scientific activity she was taken part in. Thus a person-formation (see above) that can be called subjectification – a dis-identification with the offered role as manipulator – became the consequence of the poignant experience (see chapter 4).

Whether certain actions of the students are successful in terms of learning depends on what stands fast and is used by the students when making the encountered object or feeling intelligible. In the example of Karin, we can see that the old experiences she used functioned very well, although it created a moral dilemma for her. In other cases, students do not have the relevant experiences to draw upon in order to make the situation intelligible and therefore need help to continue with the activity and in order to learn. In chapter 11, we elaborate how teachers can facilitate this.

Aspects influencing learning

Constructivist research (see for example Duit and Treagust 2003) has convincingly shown that students' earlier experience (the intrapersonal) is crucial for what they will learn in an education activity. We could also see that in the case of Karin. Her old experiences (what stand fast for her) were crucial for creating the specific relations she did and thereby for the learning outcome of the encounter with the sea urchin. Sociocultural research (see for example Wertsch 1993) has equally convincingly shown that interpersonal (social) and institutional aspects are crucial for students' learning. For example, conversations with peers are of great importance as we saw in the case of Kate, who created a counterargument in relation to what another student had said before. She thus used the knowledge and statement of another student in her argumentation. As shown in more detail in chapter 14, students often develop their arguments by elaborating on other students' views or critiquing others' reasoning. The institutional framing also influences the outcome of students' learning. Karin's inquiry on the poignant experience, for example, directly relates to the institutional framing in terms of collecting animals for a subsequent scientific investigation: It is part of the socialisation into the profession of a biologist that you learn to manipulate nature, including animals. Recently, there is a revived attention for how materiality influences learning (see for example Sörensen 2009) and it shows that the physical world is a crucial aspect influencing the learning outcomes. Karin's poignant experience includes the physical world in two ways. Firstly, it was the encounter with the sea urchin that created the experience. Secondly, the experience was physical in that it was a bodily feeling that was the object for the inquiry Karin started.

The transactional theory of learning that we elaborate in this chapter acknowledges these insights. Yet, from a transactional perspective (Dewey and Bentley 1949/1991), the cause of failure or success in bridging a gap is never connected to only one of these four aspects. These aspects do not work in isolation. Rather, it is the *interplay* between them that governs the results in terms of learning outcomes: a certain interplay creates a specific learning outcome. The four aspects influence the transactions all the way, from the disturbance to the end of the learning loops, since learning occurs when students create relations between their earlier experiences (the intrapersonal) on the one hand and the environment that they encounter on the other hand: e.g. their peers (the interpersonal), the tasks given by the teachers based on the national curriculum (the institutional) and the material objects such as a stone, a book, a sea urchin, etc.

The connection between cognitions and bodily feelings

In the following, we focus on how cognitions and feelings become intertwined when we deal with a problematic situation. When we start to actively work on a bodily feeling and a poignant experience – as Karin did through an inquiry – we can start to describe it. But before that, the feeling is vague and fuzzy. When we describe it, it is immediately transformed into something else, namely an intellectual expression (e.g. conceptualisation) of the bodily feeling. The pragmatists William James and John Dewey came up with, for their time, a revolutionary proposition, namely that *all* our knowledge, concepts, etc. have their origin in bodily feelings (James 2003; Dewey 1925/1981). Usually we think of knowledge, concepts, etc. as just intellectual – as part of reasoning. What James and Dewey are suggesting is thus that bodily feelings, which was the reason for us to think and reflect, become sedimented in the knowledge, concepts, etc. that become the result of our thinking and reflection on the bodily feeling. Such a 'marriage' between knowledge and feeling is creating what we sometimes call emotions. The most important thing, no matter what we call this marriage, is that knowledge, meaning, etc. are relational and embodied, i.e. felt in the body. Even today, this proposition can be a bit provocative, but scholars such as Mark Johnson have elaborated on it and state firmly that even

> our most abstract concepts (such as cause, necessity, freedom and God) have no meaning without some connection to felt experience.
>
> (Johnson 2007, 93)

In concrete terms, this means that if you just memorise knowledge or conceptualisations, for example the concepts of antagonism or deliberation (see chapter 8),

Figure 10.2 The model shows the start and transformation of feelings to conceptualisations (knowledge, meanings, concepts, etc.)

Source: Author's own.

you miss the important feeling that is connected to them, which makes, for example, the understanding of the political dimension in handling sustainability issues become very limited.

Figure 10.2 describes how a transaction with the surrounding world (physical as well as social) does create a bodily feeling which – when worked upon – creates conceptualisations (knowledge, etc.).

It is in concrete encounters where we create relations to the world. The specific meaning that is created depends on which relations we create between what stands fast for us and the objects and feeling we encountered. Thus, meanings are always relational, because they are created in an encounter. The meaning of the concepts that we learn are therefore both cognitively understood and bodily felt. When we for example learn what 'enemy' means, we don't just learn how to pronounce and spell it, or how to use it properly in a sentence, but also what an enemy *is*, how it feels to have an enemy or to become an enemy for someone. To get the feeling of *enemy* we have to experience situations where people become enemies or someone sharing experiences of having or becoming an enemy. Thus, in participating in an activity we learn the language *as* we learn about the world and our connections to that world (Cavell 1999). As we mentioned above, if you just memorise knowledge or conceptualisations you miss the important feeling that is connected to them, which makes the understanding of the political and the ethical dimension very limited. This is not just valid for concepts, but also the understanding of words. Imagine for example that a person in a debate would use the word 'but' in the sentence 'Your opinion is interesting, but . . .' If you don't feel the word 'but' as communicating a feeling of a quality of hesitancy or something that will go against your ideas, you would not understand the sentence. And as a consequence, you would not understand that the person was treating you as an antagonist.

Conclusion

To learn a habit is to learn a specific attentiveness: to select out of the surrounding world a fruitful environment to work with. Furthermore, to learn a habit is also to learn a specific way of coordinating with the environment in order to achieve an outcome that is expected in the activity. As sustainability problems often require creativity in order to create innovation, we want our students to learn plastic habits. Thus, as part of learning a specific coordination, artistic imagination is important.

Learning starts when our habits are disturbed – by an intellectual disruption, a change in the physical surroundings or a poignant experience. The different disturbances create problematic situations and can thus be perceived as the 'roots' for the reflection that follows and, hence, for learning. Through a short or a long learning loop, the problematic situation can be solved and the learning can result in a consolidation, enrichment or re-creation of a habit or in the creation of a new habit. From a transactional perspective, we identified four aspects that influence learning from the disturbance to the end of the learning loops: the intrapersonal, the interpersonal, the institutional and the physical. These aspects do not work in isolation. Learning outcomes depend on the interplay between them. When conceptualisation (knowledge, values, etc.) and skills are learned, the learning is not just cognitive but also bodily feelings are part of it: to understand a concept or a word is also to feel the concept and the word.

Acknowledgements

The research on which this chapter is based on was supported by Swedish Research Council (Manners of teaching about controversial sustainability issues and students learning – Grant 2017–03662).

References

Andersson, J., Garrison, J. and Östman, L. (2018) *Empirical Philosophical Investigations in Education and Embodied Experience*. Palgrave Macmillan, Cham, Switzerland.

Cavell, S. (1999) *The Claim of Reason: Wittgenstein, Skepticism, Morality and Tragedy*. Oxford University Press, Oxford.

Dewey, J. (1916/1997) *Democracy and Education. An Introduction into the Philosophy of Education*. The Free Press, New York.

Dewey, J. (1925/1981) "Experience and nature". In Boydston, J.A. ed., *John Dewey: The Later Works*, Vol. 1. Southern Illinois University Press, Carbondale.

Dewey, J. (1938/1986) "Logic: The theory of inquiry". In Boydston, J.A. ed., *The Later Works*. Vol. 12. Southern Illinois University Press, Carbondale and Edwardsville.

Dewey, J. (1938/1997) *Experience and Education*. Touchstone, New York.

Dewey, J. and Bentley, A.F. (1949/1991) *Knowing and the Known*. Southern Illinois University Press, Carbondale.

Duit, R. and Treagust, D. (2003) "Conceptual change: A powerful framework for improving science teaching and learning". *International Journal of Science Education*, 25(6): 671–688.

Gould, D. (2010) "On affect and protest". In Staiger, J., Cvetkovich, A. and Reynolds, A. eds., *Political Emotions: New Agendas in Communication*. Routledge, New York, 18–44.

James, W. (2003) *Essays in Radical Empiricism*. Dover Publications, New York.

Johnson, M. (2007) *The Meaning of the Body: Aesthetics of Human Understanding.* University of Chicago Press, Chicago.

Öhman, J. and Östman, L. (2007) "Continuity and change in moral meaning-making: A transactional approach". *Journal of Moral Education*, 36(2): 151–168.

Östman, L., Svanberg, S. and Aaro Östman, E. (2013) *From Vision to Lesson: Education for Sustainable Development in Practise.* WWF, Stockholm.

Overgaard, S. (2008) "How to analyze immediate experience: Hintikka, Husserl, and the idea of phenomenology". *Metaphilosophy*, 39(3): 282–304.

Rudsberg, K. and Öhman, J. (2015) "The role of knowledge in participatory and pluralistic approaches to ESE". *Environmental Education Research*, 21(7): 955–974.

Sörensen, E. (2009) *The Materiality of Learning: Technology and Knowledge in Educational Practice.* Cambridge University Press, New York.

Wertsch, J.V. (1993) *Voices of the Mind: A Social-Cultural Approach to Mediated Action.* Harvard University Press, Cambridge, MA.

11 A transactional theory on sustainability teaching

Teacher moves

*Leif Östman, Katrien Van Poeck
and Johan Öhman*

Introduction

In this chapter, we will present a teaching theory connected to the transactional learning theory that we introduced in chapter 3 and further elaborated on in chapter 10. An important starting point for this theory is that all environmental and sustainability teaching is directed towards the overall purposes of qualification (knowledge, skills, etc.), socialisation (values, judgements, worldviews, etc.) and person-formation (identification and subjectification). Any teaching in an environmental and sustainability education (ESE) context includes all of these broad overall ambitions, but it is always the case that one is foregrounded while the other two are in the background (see chapter 4). In chapter 10, we highlighted that learning sustainability habits (habitual actions) starts with a disturbance that encourages students to reflect. Sometimes this reflection is extended and becomes an inquiry including experimenting. In such a case, the disturbed habit would be re-created or even a new habit could be started. In this chapter we will elaborate on how a teacher can facilitate learning by performing 'teacher moves', i.e. interventions that engage the students in inquiries focusing on certain environmental and sustainability objects (for example flowers, acid rain, migration). We present a typology of teacher moves that will be exemplified in this and the subsequent chapters (12 and 13). We end the chapter by highlighting how teachers can use these moves to set a scene (learning environment) and to stage fruitful inquiries, thereby continuously managing the interplay (transactions) between intrapersonal (psychological), interpersonal (social), institutional and physical aspects that influence students' learning.

Learning: long and short loops

A short summary of our transactional theory on sustainability learning (see chapter 10) is important as a background for the subsequent presentation of the transactional sustainability teaching theory. Since we live and act most of the time in accordance to acquired habits, one of the major tasks of ESE is to encourage students to learn sustainability habits. And as a transition to a more sustainable world requires innovation and creative change, students especially need to acquire habits that allow for creativity. Any habit is built upon two elements:

attentiveness and coordination. Learning a habit is a matter of learning a specific attentiveness, i.e. the intellectual and bodily ability to select out of the surrounding world a fruitful environment to work with, putting certain objects in the world in focus. Furthermore, it is a matter of learning a specific way of coordinating ourselves with the environment in order to achieve a particular outcome. Finally, habits that allow for creativity require the learning of a specific coordination, i.e. the ability to use imagination and knowledge in order to achieve a personal vision. Such specific attentiveness and coordination require judgements. In order to select certain objects to focus on and others to neglect, and in order to include and exclude ways to coordinate with the selected object, we use values as a guiding principle. Hence, learning a habit includes qualification as well as socialisation. Since creative habits involve imagination and personal visions, it also involves person-formation.

If our acquired habits are disturbed, a 'problematic situation' arises and learning starts (see chapter 10 and Figure 11.1). Such a disturbance of a habit can take different forms: an intellectual problem (e.g. lack of understanding), a change in the physical surroundings that is not caused by human action and a poignant experience (e.g. a moral reaction or a political moment). Those different types of problematic situations can be perceived as the catalysts of reflection and thereby of learning: when a disturbance happens, the students stop doing what they are doing and start to reflect. If the problematic situation can be solved in a quite simple way, by employing knowledge, skills, etc. that were already part of an established habit, the learning path is described as a short loop. If the problem is lingering and the students need to start up an 'inquiry', a long learning loop

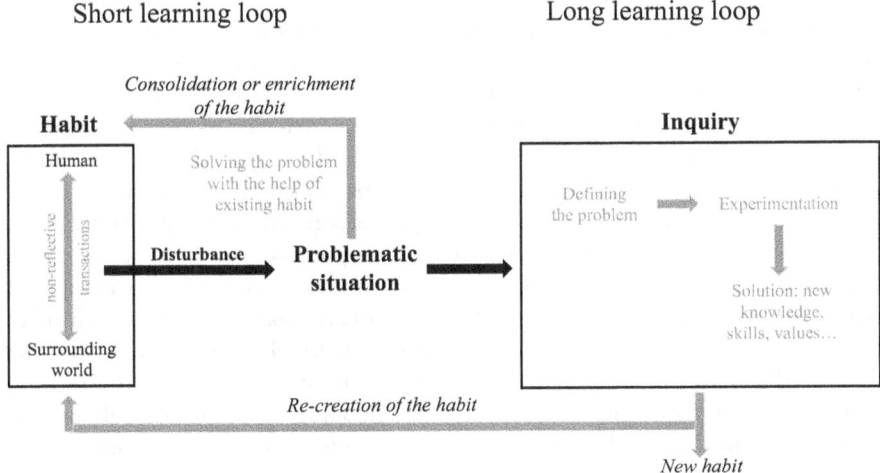

Figure 11.1 The figure shows the process from disturbance of a habit, the emergence of a problematic situation and the continuation through a short and long learning loop respectively

Source: Author's own.

is required. Through a short learning loop, the habit in use can be consolidated and enriched. A long learning loop can result in a re-creation of the habit or the start of a new habit. Figure 11.1 presents this model and shows the process from disturbance of a habit, the establishment of a problematic situation and the continuation through a short and long learning loop respectively.

Teaching: staging inquiries for the students

To be able to live a sustainable life and to create a sustainable society often requires the development of new habits, or re-construction of old habits. This process involves long learning loops. To strive for the long learning loop means that students will experience a problematic situation that requires them to start an inquiry. The question is how we as teachers can stage fruitful learning activities that lead to such long learning loops. Here the concept of *purpose* is central. Purposes, in educational contexts, are always connected to students' doing, and to communicate a purpose is to set up a learning activity, i.e. to organise students' actions to become systematically directed towards a specific learning outcome.[1] With the help of the 'organising purposes' model (Johansson and Wickman 2011), we will further clarify the meaning of purpose in teaching practice.

A purpose can be stated in the national curriculum or it can be set by the teacher in the planning of the lesson. It can also be staged within the teaching practice, and it is this meaning of purpose that is in focus here. John Dewey (1938/1997) introduced the term 'ends-in-view'. Johansson and Wickman (2011) have further developed this term in order to highlight how teachers use purposes in order to organise the students' learning activities. Dewey coined the term ends-in-view in order to differentiate between the knowledgeable teacher's comprehensive and professionally informed aims and those aims that are intelligible to students and which they can act upon. In the latter case, the expected end is in view (in reach) of the students, which means that they can start inquiries in order to reach that end. In the former case, it is not in view. If a teacher would start a lesson by communicating to the students that the purpose of the coming lessons is to gain insight into the difference between anthropocentric and eco-centric arguments in debates regarding meat consumption, they would probably not know what to do in order to reach that purpose. They cannot act upon the purpose. Johansson and Wickman (2011) call the overarching goal or aim of teaching, i.e. the purpose that only the teacher can make sense of, the 'ultimate purpose'. They use the term 'proximate purposes' for those purposes set by the teacher that can function as ends-in-view for students. Doing so, the teacher can instigate the students' self-activity, i.e. encourage them to actively engage with objects in the world (Rytzler 2017). Sometimes, however, the staged proximate purpose is not transformed to an end-in-view for students. If that happens, the teacher needs to reformulate it in order to successfully make the students understand it. Thus, teachers use proximate purposes in order to govern students' learning process towards a certain ultimate purpose. Table 11.1 describes different kinds of purposes involved in teaching and learning activities.

Table 11.1 The different kinds of purposes involved in teaching and learning activities

Ultimate purpose	Proximate purpose	Ends-in-view
The teacher's overarching and professionally informed aim, describing the expected outcome of a lesson or a series of lessons.	A purpose that the teacher communicates to the students in order to make them start an inquiry.	A purpose that students understand and can act upon in their inquiry.

In order to help the students reach an ultimate purpose (e.g. gaining insight in the difference between anthropocentric and eco-centric arguments in debates regarding meat consumption), we need to stage specific proximate purposes that will nourish the students' self-activity and initiate an inquiry (e.g. identifying arguments used in debates on meat consumption, understanding what anthropocentrism and eco-centrism means, etc.). This requires that the proximate purpose be turned into ends-in-view for the students. If that is not happening, we have to give further clarification, reorient the students' actions etc. in order to make them stage the intended inquiry. Thus, the work of a teacher has not ended when we have made a careful plan on which proximate purposes and associated inquiries need to be staged in the teaching in order for the students to reach the ultimate purpose. What we also have to do is constantly monitor how the students interpret the proximate purposes and, as a creative artist, come up with new – not planned – proximate purposes when the students run into unforeseen problems in the inquiry.

Research has shown that is very difficult for teachers to stage inquiries that result in the desired learning outcome, i.e. that fulfil the ultimate purpose (see Johansson and Wickman 2011). This is due to the contingent character of teaching, which is the case because the learning outcome is always an effect of the joint actions of both the teacher and the students. Defining ultimate and proximate purposes is of course of great help when planning lessons, but everyone with teaching experience knows that there will be many unforeseen ends-in-view that students will come up with, to which a teacher might need to respond. For example, if a student is having a poignant experience in the midst of a lesson, it opens up excellent possibilities for the teacher to engage the students in ethical or political deliberations (Garrison et al. 2015). As we will argue and illustrate below, being aware of a variety of possible 'teacher moves' can be very helpful in dealing with such challenges (see also chapter 12 and 13).

Teaching: setting the scene for the students

A learning environment consists of a selection of a limited number of objects from the surroundings. For example, if we take the students out for an ecological excursion in the woods, the surroundings consist of innumerable objects such as air, water, soil, trees, flowers and animals. Not all of them are of interest in relation to the ultimate purpose of the excursion. If one of the proximate purposes is to classify flowers in order to make an ecological investigation, then we want them to pay attention to specific flowers and to neglect the other objects in the forest. We might also bring flower guide books for the students to use. Then the books also become part of the learning environment. But it is not only physical things that can become an object of inquiry, i.e. as part of individual students' learning environment. Also knowledge, skills, ideas, values, feelings, etc. that the peers bring into the communication can become objects in an inquiry. For example, if a student in the excursion asks, 'Are we supposed to also look into the relation between the producers (for example plants) and consumers (organisms that live on the producers)?' the teacher might find that an interesting question because it points to an important ecological relation and therefore ask the rest of the group to look for this in their investigations. Thereby the relation, suggested by one student, becomes an object in each individual students' learning environment and inquiry. Not all knowledge, skills, etc. that peers bring up are relevant in relation to the proximate purpose, and the teacher thus needs to make the students pay attention to the relevant ones and neglect the others. Thus, in a teaching situation the environment consists of all the objects – physical (a leaf, a book, etc.), cognitive (knowledge, ideas, values, etc.) and bodily (feelings etc.) – that become part of students' learning activities, i.e. that become *objects of inquiry*.

It is important to offer the students a relevant and efficient learning environment in relation to the intended inquiry. This can be done by directing students' attention to certain objects, i.e. to 'set the scene' for the students' activity. These objects are the resources for the students to use in the inquiry.[2]

As teachers, we sometimes create a learning environment by collecting and bringing resources to the classroom, for example articles from a newspaper. This is what we call 'preparatory scene-setting'. At other times we just verbally bring in a new object for the students to use in the inquiry that was not planned beforehand. An example of this could be that during a discussion among students about human rights, we suggest that they look at the UN's 'Universal Declaration of Human Rights'. In this case we add something to the scene in the course of the actual teaching process. Sometimes we create a learning environment through such 'in-teaching scene-setting' by letting the students themselves search for specific resources in the surroundings. We can, for example, ask the students to gather newspaper articles on the loss of biodiversity and in a subsequent task ask them to analyse these articles regarding different views on how to increase biodiversity. When the articles are used in a subsequent inquiry that is directed towards a specific ultimate purpose, they become part of the created learning environment.

Scene-setting is a form of 'pointing out' and an important part of teaching, because here students' attention is directed, formed and shared (Rytzler 2017). The scene-setting is an offer to create relations to certain objects of the world and, through the inquiry, to do something with them. When the teacher sets

Table 11.2 The difference between the learning environment initiated by the teacher and the actual learning environment used by students

Scene-setting (preparatory and in-teaching)	Environment-in-use
Objects that the teacher tries to make the students pay attention to and use in their inquiry	Objects attended to by the students in an inquiry

a scene – a learning environment – it does not automatically imply that the students actually include it in their attention and actions. It is therefore reasonable to make a distinction between the scene-setting of the teacher and the 'environment-in-use' by the students. Table 11.2 shows this difference between the learning environment initiated by the teacher and the actual learning environment used by students.

Teacher moves

When the teacher stages an inquiry or when s/he creates a learning environment, we say that the teacher is making a move – a *teacher move*. In research, several types of teacher moves have been found (Lidar, Lundqvist and Östman 2006; Östman 2010; Rudsberg and Öhman 2010; Klaar and Öhman 2013; Van Poeck and Östman 2018). These moves have been identified by investigating the students' activity and the learning environment they are using before the transaction with the teacher in relation to the changes in the actions and the learning environment that occurs because of the teacher's intervention. A teacher's interventions can be aimed at two different didactic objectives: 'scene-setting' and 'staging an inquiry'. What is important to recognise is that sometimes staging an inquiry includes scene-setting, and vice-versa. Thus, the analytical division between the two different objectives of the interventions should be seen as a way of creating a terminology that makes it possible for us teachers to communicate about different ways of facilitating students' learning.

Within scene-setting interventions, two variants with different functions have been found: *adding* and *instructing* moves. Through an adding move, the teacher influences the learning environment by adding an object – for example a UN declaration – to be used in an on-going inquiry. Through an instructing move, the teacher sets the scene for the students – for example by encouraging them to gather newspaper articles – in order to create a learning environment for a coming inquiry. Often it takes the shape of a direct and concrete instruction on which objects to collect or create.

The interventions that stage an inquiry can be divided into two groups: moves that aim to *direct* – i.e. either to change or to confirm – the pathway that students are using in the inquiry, and moves that aim to *deepen* the students' inquiry. Within these groups, four distinct functions have been identified: *confirming*, *reorienting*, *generating* and *judging* moves. Table 11.3 gives a general overview of these different basic teacher moves.

Table 11.3 Overview of teacher moves with different basic functions and different didactic
objectives

Scene-setting	Staging an inquiry	
	Directing	Deepening
Instructing	Confirming	Generating
Adding	Reorienting	Judging

In the following, we will give examples of how we as teachers can use moves
that have a generating and judging function in order to deepen students' learning.
These examples are collected from a lesson that took place in an upper secondary
school where the students were discussing whether it is better to improve technology than change lifestyle if you want to solve environmental problems (see
Rudsberg and Öhman 2010). Although it was a discussion among the students,
the teacher participated very actively with comments and questions, which in
different ways enhanced the quality of the discussion and facilitated and deepened the students' learning in relation to the discussed topic. Some of the moves
the teacher used had a *generating function*, and the moves of this function were of
two kinds, *generalising* and *specifying*

- Generalising moves had the function of raising the conversation to a higher
  level with the aid of more general terms. An example of this move was when
  one of the students claimed: 'if in the future they succeed in making all cars
  run on air or something else that's good, fuel cells or whatever they come up
  with, that would be a definite advantage. But I think it's difficult, sure people
  could stop using their cars so much, there could be a combination . . .' The
  teacher responded to this claim by asking 'So you believe in technology?'
  When the discussion continued, the student also used the term 'technology'
  and related his example to a more general standpoint.
- Specifying moves helped the students to be more specific in their claims.
  In the continuing discussion, another student suggested that an alternative
  is to impose restrictions: 'that say that you can only shower for ten minutes
  per day, that kind of in some way changes our lifestyle to, or so that these
  environmental problems are solved'. The teacher then asked the student to
  continue and specify her contribution by asking: 'You mean a third way, to
  use legislation to . . .?' In this way, the teacher made the student aware that
  she had actually been suggesting a third option, namely legislation. The consequence of this specifying move was that the student continued and specified that legislation was indeed what she was thinking about and developed
  her argument in relation to that.

Other moves used by the teacher had a *judging function*: the *comparative* and
testing moves. These moves made the students evaluate and try out different
alternatives:

- Comparative moves mean that the teacher adds new, often opposing perspectives to the discussion. This type of move followed on from one student saying that if it is 'possible to solve all environmental problems with better technology we should of course do that'. The teacher responded to this claim by asking: 'Do you think that this is more effective than a change of lifestyle?' This comparative move made the student develop his arguments by judging it in relation to an alternative option.
- Testing moves are characterised by the teacher relating the students' statements to new circumstances. A testing move is made by the teacher after one student states that: 'We can't change our lifestyle . . .' and motivated the statement by claiming that today we dependent on technology as a part of everyday life. The teacher followed this up by saying: 'But don't you think. If the situation was to be become even more severe don't you think that people would be prepared to change their lifestyles? Say that the environmental problems became really acute, don't you think that people would be prepared to change their lifestyles then?' Here the teacher's utterance made the student inquire whether her claim still holds when circumstances change.

What these moves do in this activity is making the students learn and practice some basic skills – for example to reason in a clear and a functional way when making an argumentation – and knowledge – for example introducing terminology that makes the reasoning more general ('technology') or to make the students develop their arguments by judging it in comparison to an alternative option. It is important to notice that in this context the students are not supposed to come up with an opinion or a claim that they are strongly attached to. Here it is rather the intellectual quality of the reasoning that is pushed for (see further 'Epistemological, ethical and political moves').

Epistemological, ethical and political moves

In our teaching practice, we can make moves of different character concerning their content: epistemological, moral or political. Above we have seen examples of different *epistemological moves*. Epistemological moves are teachers' interventions that communicate which knowledge and ways of reasoning are valid and which are not in the specific activity at hand (Lidar et al. 2006). Thus, teachers use these moves to facilitate students' learning of knowledge and skills, for instance to reflect upon the knowledge basis of different common standpoints in relation to climate change. These moves aim to stimulate students' intellectual engagement in important environmental and sustainability questions.

Both ethical and political moves concern how teachers' actions facilitate students' learning when personal feelings and attachments are a crucial part of the educational situation. The difference between ethical moves and political moves is the object of students' feelings and attachments. *Ethical moves* enable students to express and share their moral experiences, to articulate ethical diversity and to deliberate on moral norms and dilemmas. Hence, the objects of students' feelings and attachments are judgements regarding the good or bad values and the right or wrong ways of treating each other and nature (see chapter 12). *Political moves*, on the other hand, are actions performed by teachers to have students raise and defend conflicting standpoints regarding sustainability issues as matters of public concern. Here, the objects are values and judgements about how we should organise society (see chapter 13).

Different transactional aspects to take into consideration

When we teachers plan our lessons and do our moves in the teaching practice, there are four aspects that influence the learning outcome (see chapters 3 and 10) that we need to take into consideration, namely:

1   the intrapersonal: the students' earlier acquired knowledge, skills, opinions, values, ideas, feelings, habits, etc.
2   the interpersonal: the social interactions between the students and teachers involved in a particular situation: dialogue, negotiation, deliberation, power relations, etc.
3   the institutional: the institutional framing through narratives, discourses, cultural artefacts, curricula, traditions, etc. that are managed by the teacher.
4   the physical: the physical objects that students encounter in the learning activity (leaves, air, book, computer, etc.).

These aspects do not work in isolation; rather they interplay: a specific interplay between them results in a specific learning outcome. From a teacher perspective, this can be described as in Figure 11.2 that describes the interplay, in the learning situation, between the intrapersonal on the one hand and the environment and the inquiry on the other.

On the left side of the figure we have the intrapersonal aspects, i.e. the knowledge, skills, values, etc. that students bring into the learning situation. On the right side we have all the objects that a teacher can use in order to stage the

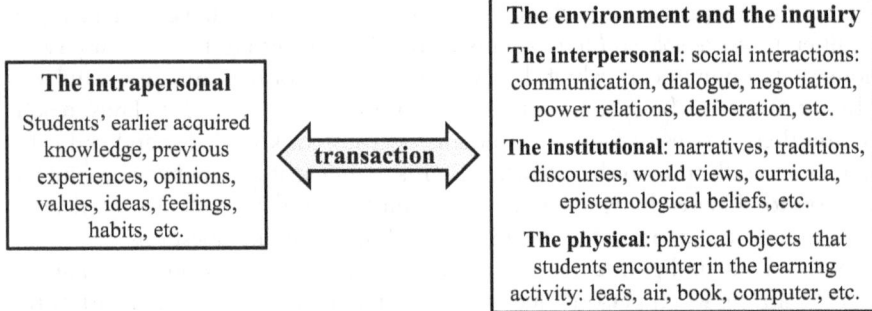

Figure 11.2 The interplay between the intrapersonal, the environment and the inquiry
Source: Author's own.

inquiry and to set the scene, i.e. the environment for the students to use in the inquiry. Here we find the interpersonal aspects connected to questions such as 'Should the students work individually or in groups?' and 'How should the groups be composed?' The institutional aspects are also situated on the right side of the figure. Here, the teacher has to decide which concepts, models, ways of reasoning coming from disciplines, arts etc. the students will be offered as resources to work with in the inquiry. Furthermore, the inquiry process that the teacher privileges is also part of the institutional aspects, since it is almost always connected to purposes put forward in the national or local curriculum. The physical aspects, finally, concern the question of which physical objects should be part of the students' inquiry.

As a teacher, we constantly have to manage the interplay between these different aspects. We have to take into consideration students' earlier experiences when deciding on which environment to set and what inquiry to stage. This is crucial because we know from research[3] that the objects and the task (inquiry) are only intelligible to the students if they can use these previous experiences in order make the objects and tasks meaningful. Only then will the proximate purpose become an end-in-view for the students and the scene set by the teacher become the environment in use for the students (see Table 11.2). But we also know from research[4] that decisions concerning, for example, which physical objects, which concepts and which peers to work with have direct consequences for the learning outcome. For instance, if we offer students a wall map or a projection of a map on a smart board to work with in an inquiry in order to find out in which direction the water flows in German rivers, many students will answer that it flows to the south. The reason for this is the fact that water flows from higher to lower altitudes and here that would be from the top (north) to the bottom (south). Thus, these maps direct the students' reasoning into wrong directions. If the students would use a relief globe, their way of reasoning would be different and correct, because then they can see how the direction from higher to lower altitude depends on the formation of the mountains. Hence, it is not enough to take into account students' earlier acquired knowledge, skills, experiences, etc. but also to consider whether the staged environment will lead their attention in the right direction. Important to recognise is that the intrapersonal on the one hand and the environment and inquiry (interpersonal, institutional, physical) on the other cannot be taken into account separately. They need to be considered as in interplay between each other. If the students involved in the inquiry mentioned above, for instance, would have seen a TV programme just before that explained the flow of water in the landscape, they would be prepared in such a way that the artefact would not guide them in a wrong learning direction. Thus, the intrapersonal aspect of the interplay (previous knowledge) would be different and allow the students to use a (topographic) wall map correctly in order to answer the question. The conclusion is that in order to set a scene and stage an inquiry that can be made intelligible by the students and at the same time steer their learning in the right direction, one needs to think transactionally, i.e. to recognise the interplay between intrapersonal, interpersonal, institutional and physical aspects.

Table 11.4 An overview of different types of teacher moves

Didactic objective of the moves	Function of the moves		Some examples of specific moves		
			Epistemological moves	Ethical moves	Political moves
Scene-setting	**Instructing** **Adding**		• Norm-installing move (see chapter 13)		
Staging an inquiry	**Directing**	**Confirming** **Reorienting**	• Reinstating move • Rationalising move (see chapter 13)		
	Deepening	**Generating**	• Generalising move • Specifying move (see 'Teacher moves')	• Clarifying ethical move • Articulating ethical move (see chapter 12)	
		Judging	• Comparative move • Testing move (see 'Teacher moves')	• Evaluating ethical move • Testing ethical move • Controversy-creating ethical move • Hierarchizing ethical move (see chapter 12)	• Controversy-creating move • Hierarchizing move • Excluding-including move (see chapter 13)

Summary

In this chapter, we have described teaching as a matter of staging inquiries for the students (through ultimate and proximate purposes and ends-in-view) and setting the scene (creating a learning environment). A teacher can stage an inquiry or create a learning environment by making teacher moves. There are several types of teacher moves. We distinguished scene-setting moves and moves to stage an inquiry. Within the scene-setting moves, there are two variants with a different function: adding and instructing moves. Moves that stage an inquiry can be divided into two groups: directing and deepening moves with four different types of functions: confirming, reorienting, generating and judging functions. In each of these categories, teachers can make epistemological, moral and political moves. Table 11.4 gives an overview of the different types of teacher moves. Some of the moves we have illustrated above, while others will be elaborated on in chapters 12 and 13.

As teachers, we can use these moves to set the scene in a functional and relevant way and to stage fruitful inquiries. Being aware of the wide variety of possible teacher moves is vital to successfully manage the interplay between intrapersonal, interpersonal, institutional and physical aspects that influence students' learning.

Acknowledgements

The research on which this chapter is based on was supported by Swedish Research Council (Manners of teaching about controversial sustainability issues and students learning – Grant 2017–03662).

Notes

1 It is important to emphasise here that systematically directing students' actions towards 'specific learning outcomes' should not be understood in a strictly instrumental way, as only a matter of learning the 'proper' knowledge or acquiring predetermined, 'sustainable' values and attitudes. Purposes can also involve organising students' actions towards learning to explore different standpoints or taking a personal stand, for example.
2 It is worth noticing that there are differences between schools and universities around the world regarding the educational surroundings, which creates different pre-conditions for teachers to set a fruitful learning environment for the students.
3 Constructivist research has convincingly shown that students' earlier experience (the intrapersonal) is crucial for what they learn in an education activity (see chapter 10).
4 This has been shown by sociocultural research (see chapter 10).

References

Dewey, J. (1938/1997) *Experience and Education*. Touchstone, New York.
Garrison, J., Östman, L. and Håkansson, M. (2015) "The creative use of companion values in environmental education and education for sustainable development: Exploring the educative moment". *Environmental Education Research*, 21(2): 183–204.
Johansson, A.-M. and Wickman, P.-O. (2011) "A pragmatist understanding of learning progressions". In Hudson, B. and Meyer, M.A. eds., *Beyond Fragmentation: Didactics,*

Learning and Teaching in Europe. Barbara Budrich Publishers, Leverkusen, Germany, 47–59.

Klaar, S. and Öhman, J. (2013) "Doing, knowing, caring and feeling: Exploring relations between nature-oriented teaching and preschool children's learning". *International Journal of Early Years Education*, 22(1): 37–58.

Lidar, M., Lundqvist, E. and Östman, L. (2006) "Teaching and learning in the science classroom: The interplay between teachers epistemological moves and students' practical epistemology". *Science Education*, 90(1): 148–163.

Östman, L. (2010) "Education for sustainable development and normativity: A transactional analysis of moral meaning-making and companion meanings in classroom communication". *Environmental Education Research*, 16(1): 75–93.

Rudsberg, K. and Öhman, J. (2010) "Pluralism in practice: Experiences from Swedish evaluation, school development and research". *Environmental Education Research*, 16(1): 115–131.

Rytzler, J. (2017) *Teaching as Attention Formation: A Relational Approach to Teaching and Attentions.* Mälardalen University Doctoral Dissertation, 221.

Van Poeck, K. and Östman, L. (2018) "Creating space for 'the political' in environmental and sustainability education practice: A political move analysis of educators' actions". *Environmental Education Research*, 24(9): 1406–1423.

12 Ethical moves

How teachers can open up a space
for articulating moral reactions and
deliberating on ethical opinions
regarding sustainability issues

Katrien Van Poeck, Leif Östman and Johan Öhman

Introduction

This chapter focuses on teachers' influence on students' ethical and moral learning and, in particular, on how teachers can promote students' growth as moral subjects in environmental and sustainability education (ESE) practice. By performing specific 'moves' (see chapter 11), a teacher can affect how the ethical dimension of sustainability issues (see chapter 6) is experienced in educational practice. Teacher moves are actions – practical and/or conversational – of a teacher that bring about a change or enforcement of the direction of students' learning. These actions affect what is taken into account and what is not and thereby govern the learning process in a certain direction. From the perspectives of sociocultural learning theory, this process is called 'privileging' (Wertsch 1993). As a result of teacher moves, students consider certain meanings, questions, artefacts, etc. as reasonable, while others – although fully conceivable – are ignored or disregarded (Östman 2010). Teacher moves have been distinguished in different domains. 'Epistemological moves' are teachers' interventions that communicate which knowledge and which ways of creating knowledge are valid and which are not (Lidar et al. 2006). 'Political moves', then, are actions performed by teachers to have students raise and defend conflicting standpoints regarding sustainability issues as matters of public concern (see chapter 13). In this chapter, we introduce 'ethical moves', i.e. actions that we can perform in our teaching to open up a space for articulating moral reactions and deliberating on ethical opinions. By presenting, illustrating and discussing a variety of possible teacher moves, our aim is to offer tools to reflect upon and consider different possible ways of acting in view of students' ethical and moral development in relation to sustainability issues.

Ethical and moral learning

Ethical moves focus on the ethical dimension of sustainability issues, i.e. the good values that people find desirable and the right actions that reflect these values. We are here dealing with statements and judgements of what is right or wrong, good or bad. Such statements and judgements can be handled and experienced in varied ways in education. Chapter 6 presents and describes a typology of the different ways in which morals and ethics can appear in educational practice: the ethical tendency (Öhman and Östman 2008). It distinguishes between three different ways in which attitudes to the right and the good are communicated: moral reactions, norms for correct behaviour and ethical reflections. 'Moral reactions' are those situations when we, without any previous consideration or reflection, take spontaneous responsibility for another being. This can happen, for example, in response to seeing someone or something that is in need or that is treated badly. Moral reactions are thus unpremeditated reactions to situations in which we are personally and emotionally affected and experience a deep, bodily feeling of care or disgust. 'Norms for correct behaviour' are a very different way to experience opinions about the right and the good. Here, it is about situations in which social rules that prescribe the morally correct way of acting in certain kinds of situations are communicated to us by the way in which authorities and peers actively respond to our actions. The norms indicate a common opinion and can therefore often be formulated in terms of rules for the way we are supposed to act. Following norms thus has to do with what knowing what we *should* do to fulfil the expectations of our fellow beings. 'Ethical reflections' are those situations in which we start a critical inquiry on good values and the right way of acting. It is about formulating and comparing systematic and rational arguments for how to handle certain moral issues. It is a rather 'distanced' way of engaging with the ethical dimension of sustainability issues as, while reflecting, we are usually not in an immediate situation of needing to decide how to act but focus on general ethical principles that human beings ought to follow. In some cases, however, we deal with arguments about the right way of acting in situations where we are expected to take a stand on a particular ethical issue and explain and defend our personal standpoints. We call this specific situation 'ethical deliberation'.[1] It often involves strong feelings and personally felt attachments as it does in moral reactions.

Obviously, various ways of dealing with the ethical dimension of sustainability issues bring about very diverse educational practices. Teachers' moves are one way of influencing this besides, for instance, students' behaviour and the design and purpose of the lesson. In chapter 5, we have argued in favour of a pluralistic approach to ESE and highlighted the importance to involve diverse interests, perspectives, values, worldviews, etc. From this point of view, it is vital to create opportunities for students to encounter situations in which different expressions of the right and the good can be connected to sustainability issues and to encourage them to articulate their moral reactions, ethical standpoints and personal beliefs and feelings. As teachers we can do so by using 'ethical moves': actions that can facilitate students' learning in ethical deliberation as well as in articulating and sharing moral reactions and dilemmas that can become a consequence of a moral reaction.

Recognising opportunities for learning through moral reactions and moral dilemmas

In this book, we build on John Dewey's perspective that morals are not something innate or fixed, but something that we learn through a continuous process throughout life, i.e. that learning is driven by experiencing moral situations that make us reflect on responsibilities and concerns (see chapter 7). Hence, a crucial teaching principle is to start our teaching in students' personally felt experiences of moral situations that emerge in real life. It is therefore important for us as teachers to recognise situations where students express personally felt experiences of moral situation and to grasp these learning opportunities by performing well-considered teacher moves in order to nourish authentic ethical inquiries. Below, we discuss two examples of education activities where such an opportunity turned up but where the teacher did not use it for such purpose. We want to stress here that this discussion should not be regarded as criticism of the teachers' practices. They may have had very good reasons for not performing ethical moves there and then, for instance a lack of time, the students' age or the wish to avoid conflicts in the group. We discuss the examples, however, as they can provide insight into the emergence of authentic moral experiences in the classroom and how moral moves can be employed to turn them into fruitful starting points for ethical and moral learning.

Facilitating learning from moral reactions

The following example comes from an observation of an excursion in a nature reserve for a group of secondary school students (Van Poeck et al. 2016). The teacher, a nature guide, takes them for a walk through the reserve and draws their attention to several things they encounter such as plants, animals, specificities of the landscape, etc. When they see a group of Canada geese, he explains how they try to protect the nature reserve against the negative impact of these exotic species:

Teacher: What you can see there, for instance,. . . is a number of Canada geese. They are very aggressive breeding birds over here. And they chase away the native breeding birds. Canada geese are an exotic species, chasing off the pairs of black-tailed godwits that come here from Africa to breed, for instance. I said they are an exotic species and in this natural reserve, where all flora and fauna are protected, it has been decided to cull them, to control their number, at the moment there are several out there in those meadows. But since shooting is not allowed in natural reserves . . . they have opted to destroy the nests to reduce their population a bit . . . They addle the eggs and put them back in the nest or they puncture them and put them back in the nest. And mother goose simply continues brooding but those eggs won't hatch.

Inez: Oooh.

Teacher: And after a while, after quite a number of days following her usual breeding period, the goose realises that, oh dear, something's wrong here, but by then the breeding season no longer allows her to nest again.

Inez: Oooh.

Teacher: So those are destroyed.

Inez: That's how you control, they put the egg there, but what about that Canada goose?

Teacher: Well, we're not killing the Canada geese, those that are there, can stay, but we're controlling their numbers. And even so, in winter we've got concentrations of up to hundreds of Canada geese that elsewhere in Flanders . . . there they are shot, sometimes by the hundreds.

Inez: Oooh.

Teacher: Not that I'm against that, and I can even accept their being shot in a particular way. But what I do object to is that, once those . . . animals have been shot, they bring in a crane and a bulldozer, dig a hole and shove them into it. That's something I do . . . that's unacceptable from an ethical point of view. If you kill animals, you should show them some respect, for instance by using them for food . . . That's a personal thing. Please come along. [They return to the nature education centre and the excursion is over.]

What we see here is a student that repeatedly expresses some form of being emotionally affected by the teacher's explanation. Inez's non-verbal communication ('Oooh') and the utterance 'but what about that Canada goose?' can be seen as a moral reaction. It is a spontaneous, unpremeditated response to a situation in which she is emotionally affected and experiences a feeling of care, in this case for the Canada geese and their nests. The teacher – perhaps due to time pressure – ignored the repeated non-verbal expressions of the student's emotional affectedness and responded to the question 'what about that Canada goose?' by explaining the policy in this reserve – 'we're not killing the Canada geese' – comparing it to other policies – 'elsewhere in Flanders . . . there they are shot, sometimes by the hundred' – and expressing his personal ethical standpoint regarding this issue. Another possible reaction could have been to perform a 'controversy-creating ethical move', encouraging the students to judge his ethical standpoint and argue for their own. Or an 'articulating ethical move', encouraging Inez to articulate her reaction and voice more explicitly the opinions about the good values and right actions connected to it. For her, this could have been a valuable opportunity to reflect upon, make meaning of and, hence, learn from her personal and strongly felt experience. It could also, for the whole class, have functioned as a fruitful starting point for an engaged ethical discussion about questions such as whether native species have higher value than exotic species, whether the life of unborn birds should be protected, whether it is right to kill certain animals for the sake of others, etc. This could offer students the opportunity to learn to respect the deeply personal, moral feelings that people show in different situations, even though these feelings may not always be possible to explain or defend by rational argument.

Facilitating learning from moral dilemmas

The next example comes from an observation of a workshop for primary school students (mixed age, ranging from 6 to 12 years old) aimed at creating animation films about 'The city of my dreams'. At the start of the workshop, the teacher asks all the children to present themselves and to talk about what makes them happy, what makes them angry, what they wish for the future, what they would want to change in the world, etc. When it is his turn to answer these questions, Wannes, a 7-year-old boy, says:

Wannes: What I would like is that people leave nature alone and that the hunters would not shoot the animals.
Rens: Do you eat meat?
Wannes: Yes, but when I'm twenty I will become a vegetarian.
Teacher: But now, as a kid, you still need to eat meat don't you.
Isaak   [pointing to his friend]: He is a vegetarian.
Teacher: Yes, but you can only do that if your parents really know very well what to do. Otherwise it can be dangerous.
Wannes: My mom and dad do know that very well, 'cause my mom is a vegetarian. But I am not yet 'cause I like to eat meat. And what I would like is that the whole world would be made of sweets. And that you would not die.

Rens's intervention – 'Do you eat meat?' – opened up an opportunity for an ethical deliberation about Wannes's concern about killing animals: Can one defend the standpoint that it is wrong to shoot animals and, at the same time, consume killed animals? Wannes's immediate response shows that he has, indeed, considered that question and come up with the compromise to become a vegetarian in the future. The teacher, however, steers the discussion away from value conflicts and possible moral dilemmas. Instead of urging Wannes to experience the dilemma implied in the conflict between his opinion that it is wrong to kill animals and his choice to eat meat, she brings in an argument that legitimises the compromise – 'But now, as a kid, you still need to eat meat don't you'. When Isaak introduces a counterargument – 'He is a vegetarian' – she continues with the same strategy – 'Yes, but you can only do that if your parents really know very well what to do. Otherwise it can be dangerous'. Interestingly, Wannes himself counteracts the validity of this argument – 'My mom and dad do know that very well, 'cause my mom is a vegetarian' – and explains his decision by referring to another personal concern – 'I like to eat meat' – that is in conflict with his ethical standpoint that it is wrong to kill animals. Then he starts to talk about something else and the teacher does not return to the initial topic of debate. Perhaps she wanted to avoid conflicts in the group or considered (some of) the children too young for such ethical deliberation. In case that we, through our teaching, would want to help students to experience a moral dilemma and discover (new) solutions in doing so, we can perform a 'hierarchizing ethical move'. Here, for instance, this could be to ask Wannes a question like: 'So, for you, enjoying the taste of meat is for the time being more important than your worries about killing animals?' Or, in order not to put too much pressure on one student, to open up this discussion with the whole group, for instance with a values-clarification exercise (see chapters 6, 18 and 19). Below, we discuss an example of such a discussion.

Teaching and learning ethical deliberation

In this section, we illustrate some ethical moves and their impact on students' learning with an example that we observed during a lesson in an upper secondary school (Öhman and Östman 2007). It concerns an exercise aimed at ethical deliberation, which is a crucial teaching strategy because it is about making students take a personally engaged standpoint. Thus, it goes further than engaging students in an intellectual cognitive exercise, since it also involves values and feelings. The teacher made ethically significant statements about sustainability issues, such as 'Man and animals are of equal value; it is always wrong to kill an animal!', 'No life must be sacrificed as a result of environmental pollution!' and 'Everybody has the right to not sort their garbage!' The students were asked to take a stand in relation to each statement and, in the ensuing discussion, explain and defend their individual standpoints. At the start of the exercise, the students are sitting on chairs placed in a ring. If they do not agree with the statement of the teacher they shall stand up, if they agree they remain seated. In the discussion, one student responded to the teacher's statement that 'Man and animals are of equal value; it is always wrong to kill an animal!' in the following way:

Martin: Well, you have to survive!
Teacher: You don't have to eat meat to survive, do you?
David: Well, we are not at the top of the food chain for no reason. Nature works in its own way and it is because we know how to kill them that we are on the top of the food chain, it is the law of nature. The survival of the fittest . . .
Robin: Well, that is one thing, I do agree with that. The wrong thing is turning it into mass production, I mean to produce animals just to slaughter them. I can't change seats because I eat meat, and then I would be contradicting myself but I can understand the reasoning anyway, because I think it is a bit wrong to produce animals for this purpose. I can understand only killing animals in the wild, but there would not be enough animals for that. Because we want meat to eat, we have created some sort of mass production, so maybe it is wrong to do that, then maybe we should discuss in what way we treat animals. And there I think it is important, 'cause if you in some way declared equality between man and animals then you will come to the kingdom of plants at the next level, 'cause there are some who claim that they are just as equal as man and you should only eat windfalls or just eat a certain part. In this way it would always lead to a further debate. I think that animals hold the same value as man anyway, so I can agree on the first. But I will never stop eating meat so therefore I must agree on that we have the right to, right and right, but we do it because of . . .
Jeff: Well, I totally agree with [Boy 3], the only thing is that I don't eat meat, that's why I'm sitting . . .
Teacher: Is it from the point of view that they have the same value . . .?

Jeff: Well, I don't know if they hold exactly the same value as we do, but it is wrong to kill anything, it always is. But we have to kill to survive, but it is the overconsumption I'm against. But now I'm just sitting here because I don't eat meat and then I feel it's wrong not to sit here . . . I was thinking, it would be hard to take a gun and shoot a sheep, just like that, because the sheep isn't worth as much as I am, so then I can shoot it. It's a psychological thing, just like that, I have difficulty with that. If I would sit there in the Middle Ages with nothing to eat I would easily kill the sheep, so it is a luxury problem too . . .

The value statement that the teacher makes as a start of the inquiry – 'Man and animals are of equal value; it is always wrong to kill an animal!' – has the function that the students start to make judgements. Thus, the students are encouraged to take a standpoint and furthermore to argue for it. This particular way of encouraging judgement is called an 'evaluating ethical move', because the students have to decide whether they agree or not. This is also what they do. The teacher's first intervention (in response to Martin) – 'You don't have to eat meat to survive, do you?' – has a judging function (see chapter 11). The teacher creates this by questioning Martin's argument – 'Well, you have to survive!' This specific way of making the students to judge is called a 'testing move'. It makes David respond by another fact-based argument. Robin and Jeff continue the argumentation by taking a moral standpoint and argue for it with the help of facts *and* values. It is interesting to notice the argumentation developed by Robin where he is caught up in a dilemma: he agrees with the first part of the statement – 'Man and animals are of equal value' – but he does not agree with the second part – 'it is always wrong to kill an animal' – since he eats meat. He tries to argue for his non-agreement, but the dilemma becomes visible when he cannot find support for his claim that humans have the right to kill animals. The second intervention of the teacher – 'Is it from the point of view that they have the same value?' – has a generating function: it deepens and qualifies Jeff's argumentation. This specific type of generating function is called a 'clarifying ethical move' since it makes the student elaborate on the argumentation he developed before the teacher made the intervention.

Conclusion: 'moving' towards students' growth as moral subjects

We have in this chapter described and discussed several ethical moves, i.e. actions that we can perform in our teaching to open up a space for articulating and elaborating on moral reactions and dilemmas as well as deliberating on ethical standpoints. The moves are summarised in Table 12.1.

By conducting ethical moves, a teacher can facilitate students' growth as moral subjects, i.e. to develop, based on understanding and feelings, moral standpoints as well as ethical reflections and argumentations. By articulating and making meaning of moral reactions and dilemmas, by raising and defending different ethical opinions, students can learn for example the following: They increase

Table 12.1 Examples of ethical moves and their function

Functions	By these ethical moves	Description
Generating	Clarifying ethical move	A teacher's action that makes the students generate a deepened way of arguing by clarifying the ethical claim or the arguments made for the claim
	Articulating ethical move	A teacher's action that makes the students generate an articulation of their ethical standpoints and the argumentation for it
Judging	Evaluating ethical move	A teacher's action that makes the students ethically judge through evaluating an ethical standpoint
	Testing ethical move	A teacher's action that makes the students ethically judge through testing the validity of an argument
	Controversy-creating ethical move	A teacher's action that makes the students ethically judge through staging a controversy
	Hierarchizing ethical move	A teacher's action that makes students prioritise amongst different values and thus take a stand on which concerns about the good values and right actions take precedence and which must give way

their communicative ability in addressing ethical and moral issues; they learn to formulate valid arguments for ethical standpoints and to consider other people's arguments; they learn to respect the deeply personal, moral feelings that people show in different situations; they expand their awareness of different moral reactions and their ability to understand the various existential questions; they learn that human actions are not entirely based on rational decision-making; they learn to critically reflect on moral norms, evaluate different alternatives and, perhaps, influence norms (see further chapters 6 and 7). All this is very much in line with the pluralistic approach to ESE (see chapter 5). Ethical differences are not eliminated or denied but, on the contrary, made manifest, explored and used as a fruitful driver for ethical and moral learning.

Acknowledgements

The research on which this chapter is based was supported by Ghent University's special research fund (research project 'Urban sustainability transitions as spaces for experiential learning' – Grant BOFSTA2016001001) and FORMAS (research project 'Wicked problems and educative spaces for urban sustainability transition' – Grant 2016–00992_3).

Note

1 Note that 'ethical deliberation' is not part of the ethical tendency typology described in chapter 6. In the ethical tendency, deliberation is included in 'ethical reflection'. However, as this chapter deals with 'ethical moves' that a teacher can perform to open

up a space for articulating moral reactions and deliberating on ethical opinions, it is important to distinguish between forms of ethical reflection that are a more 'distanced' way of systematic and rational inquiry of how to handle certain moral issues on the one hand, and situations of deliberation where we take a stand and defend our personal, often emotionally invested standpoints on the other.

References

Lidar, M., Lundquist, E. and Östman, L. (2006) "Teaching and learning in the science classroom". *Science Education*, 90(1): 148–163.

Öhman, J. and Östman, L. (2007) "Continuity and change in moral meaning-making: A transactional approach". *Journal of Moral Education*, 36(2): 151–168.

Öhman, J. and Östman, L. (2008) "Clarifying the ethical tendency in education for sustainable development practice: A Wittgenstein-inspired approach". *Canadian Journal of Environmental Education*, 13(1): 57–72.

Östman, L. (2010) "Education for sustainable development and normativity: A transactional analysis of moral meaning-making and companion meanings in classroom communication". *Environmental Education Research*, 16(1): 75–93.

Van Poeck, K., Goeminne, G. and Vandenabeele, J. (2016) "Revisiting the democratic paradox of environmental and sustainability education: Sustainability issues as matters of concern". *Environmental Education Research*, 22(6): 1–21.

Wertsch, J.V. (1993) *Voices of the Mind: A Sociocultural Approach to Mediated Action*. Harvard University Press, Cambridge, MA.

13 Political moves

How teachers can open up for and handle poignant experiences of the conflictual aspects of sustainability issues

Katrien Van Poeck and Leif Östman

Introduction

This chapter focuses on teachers' influence on students' learning regarding the political dimension of sustainable development. As we have elaborated in chapter 8, this encompasses a wide variety of situations in which teachers and students engage with the question of how to organise society sustainably, acknowledging that this inevitably requires judgements, prioritisations and decision-making about different and competing alternatives (see further later). By performing specific 'teacher moves' – i.e. interventions made by a teacher to govern the students' learning in a certain direction (see chapter 11) – we can affect how the political is experienced in educational practice. That is, through specific types of teacher moves, we can steer to a certain extent how the political dimension of sustainability issues will appear in education activities, knowing that this has a strong influence on the learning opportunities offered for the students (see chapter 8). For instance, engaging students in 'political reflections' will make them rationally reflect on how we organise social life as well as on the different, competing opinions, judgement and prioritisations regarding what is a good, right, just, sustainable, etc. ways to organise our society. Such reflections do not require the students to take a personal stand and defend it. It is a more 'distanced' way of systematic and rational inquiry through which the students will gain new insights and understanding by approaching the object of their inquiry as an intellectual problem (see chapter 10). When the political dimension of ESE appears in the form of 'conflict-oriented deliberation', very different conditions for learning are created. Here, the students are engaged in raising and defending opposing and contesting perspectives, without the aim that a consensus has to be reached. This offers them opportunities to learn from a first-hand experience of the conflictual aspects of sustainability issues and from the strong feelings involved when one is personally attached to a specific opinion, concern, etc.

Below, we will pay particular attention to the latter kind of situations. As argued in chapter 10, the poignant experiences that can emerge when the conflictual aspects of sustainability issues are deeply and personally felt are an important source for learning. We introduce the notion of 'political moves' as a specific type of moves that teachers can use in order to give shape to an environmental and sustainability education (ESE) practice as a 'conflict-oriented political deliberation', in which students raise and defend conflicting standpoints regarding matters of public concern. This offers them the opportunity to experience judgement in situations that are (a) not indifferent to them (something that matters is at stake), (b) where the prioritisations and decisions at stake regard mutually exclusive alternatives (choosing for one implies excluding the other) and (c) are taken in a context of 'undecidability', i.e. without universal ethical or rational foundation that can uncontestably steer judgement, prioritisations and decision-making. Thus, ESE practice in the form of conflict-oriented deliberation creates a fertile ground for learning from a first-hand experience of the conflictual aspects of sustainability issues. We will illustrate and discuss this, drawing on an example of a teacher that performs political moves, and subsequently highlight how this creates very specific learning opportunities by contrasting it with another example, one where the teacher closes down the space for the learners to involve their own, personally engaged concerns and to experience the conflictual in the judgements, prioritisations and decisions at stake.

Besides teachers' actions also other aspects, such as students' behaviour and the design and purpose of the lesson, have a major influence on whether and how conflict-oriented deliberation can emerge and be handled in education practice. However, presenting, illustrating and discussing a variety of possible teacher moves can support teachers to design and perform conflict-oriented deliberation as well as to deal with poignant experiences when these (unexpectedly) occur in the classroom. It offers us tools to reflect upon and consider different possible ways of acting in relation to our teaching purpose.

Political moves

We introduced the concept of 'teacher moves' to describe how teachers' actions affect students' learning by influencing what is taken into account and what is not and thereby governing the learning process in a certain direction. Epistemological moves are tools for epistemologically oriented practices focusing on the learning of knowledge by communicating which knowledge and ways of reasoning are valid and which are not (see chapter 11). Ethical moves are actions of a teacher that open up a space for articulating moral reactions and deliberating on ethical opinions (see chapter 12). Political moves (Van Poeck and Östman 2018), then, are focused on giving the students an opportunity to learn from a first-hand experience of conflicting standpoints, concerns, etc. regarding how to organise society sustainably and from the strong feelings involved in such experiences.

As teachers, we can either open up or close down a space for raising and defending conflicting standpoints regarding how to organise society in the face of sustainability challenges. The typology 'political tendency' (chapter 8) is a classification of different possible ways in which the political dimension of ESE can be handled and experienced in practice. The teacher's interventions are one of the elements that affect this. For instance, with our teacher moves we can influence whether students engage in a systematic and rational inquiry of different standpoints regarding how to organise society without involving their own personally engaged concerns (i.e. political reflection) or rather in raising and defending personal opinions and taking a stand in relation to different, conflicting alternatives (i.e. political deliberation). Our teacher moves can also orient such deliberation towards an a priori specified desired outcome (i.e. normative deliberation), towards resolving all divergence into one shared standpoint (i.e. consensus-oriented deliberation) or towards raising and defending opposite and contesting perspectives (i.e. conflict-oriented deliberation). Often, students have already developed strongly felt concerns and opinions regarding specific sustainability issues that they bring with them into the classroom and mobilise while engaging in political deliberation. Sometimes, however, they suddenly and unexpectedly experience such strongly felt concerns and opinions in the educational activity. When the latter is happening, the students experience a political moment (Håkansson and Östman 2018 – see also chapter 8). A political moment is a situation where a student unexpectedly has a strong, poignant experience of the conflict involved in the judgements, prioritisations and decisions at stake. The feelings of antagonism that this can bring about are connected to the fact that any prioritisation or decision requires the inclusion and exclusion of opinions and concerns. As the etymology of the word reveals, a decision inevitably involves exclusion. The term stems from the Latin word 'decidere', which is constituted by 'de-' ('off') and 'caedere' ('to cut'). Hence, 'de-ciding' is not merely a matter of choosing, of making up one's mind, but it is about settling a dispute by making a distinction between what will be included and excluded ('cut off'). Thus, when people enter into a political moment, they have a strong experience that the prioritisation and decision at stake implies an inevitable demarcation between which emotionally invested commitments and concerns will be taken care of and which must give way. There is thus always a risk that one's own personally engaged commitments and concerns will be excluded.

It is not difficult to imagine, for example, that in a discussion about whether governments should instal flight quota (a maximum number of flights/air miles per citizen per year) for all citizens in order to protect the climate, a diversity of personal standpoints, commitments and concerns would be mobilised. It might also happen that people experience a political moment giving rise to a political commitment in the issue that they did not have beforehand. Many of the standpoints, commitments and concerns would be impossible to reconcile: Should we protect the habitat of millions of people threatened by rising sea levels or individuals' freedom to explore distant cultures and landscapes? Should we take care of those people that, already today, are suffering from increasing drought, forest fires, hurricanes, etc. and therefore deny others' right to study or work abroad? Etc. Such judgements about what should be prioritised and what must give way are not merely the subject of rational, cognitive reflection. They involve a deeply felt, personal experience because something that is highly valued is at stake. Research has shown that if ESE practices take the form of conflict-oriented deliberation, occasions for poignant experiences are more likely to open up (Van Poeck et al. 2016; Van Poeck and Östman 2018; Håkansson et al. 2018;

Håkansson and Östman 2018). By inquiring into a poignant experience, students can learn and make meaning of the political in relation to the sustainability issues in focus (see chapter 10: a long learning loop starting from a poignant experience – we will return to this in the conclusion).

Before we exemplify different political moves, we want to highlight that learning to make meaning of the political requires a focus on matters of public concern, i.e. questions regarding how to organise society in the face of problems that involve a diversity of conflicting attachments and concerns. Moreover, it is important that the controversy is not played out in the rational or moral register – as a matter of 'true or false' or 'good or bad/right or wrong' – but in political terms, i.e. as a struggle between different legitimate standpoints on how to 'de-cide', on what should be included and excluded (Mouffe 2005; Todd 2010).

'Moving' towards conflict-oriented deliberation

As argued, by using political moves in our teaching practice, we can create opportunities for students to have a poignant experience of the political or to articulate their strongly felt concerns and opinions through orienting their activities towards raising and defending conflicting standpoints and perspectives on sustainability issues. We will illustrate this with some examples of a guided tour of an ecological farm (community-supported agriculture) for a group of university students in bioscience engineering (Van Poeck and Östman 2018). The tour was guided by the farmer, so we focus on the teacher moves he performs. As the examples below show, he steers the activity towards a conflict-oriented deliberation by performing a combination of several judging moves (i.e. moves that deepen the students' learning by encouraging them to make a judgement – see chapter 11): controversy-creating moves, hierarchizing moves and excluding-including moves.

If you want, as a teacher, to avoid that your students' discussion remains within the logic of a consensus-oriented conversation, you can perform 'controversy-creating moves'. The farmer often used these moves to involve the students in the discussion. Throughout the conversation, he repeatedly called on the students to object to his explanations and opinions if they disagreed with him, e.g., 'But tell me if I . . . say something that doesn't make sense OK? Because this is just my subjective story'. Or, 'But you can object to me, right, I really want it to be a dialogue'. These interventions allowed him to make the students privilege an argumentative discourse: taking a stand, defending one's position with arguments, contest the farmer's and each other's points of view with counterarguments, etc. Below we present an example of how a controversy-creating move ('Well, you can say if it's true or not OK?') provokes a student to react on the famer's standpoint by contesting it:

Farmer: So it's [profit] that now has the upper hand in agriculture. All the farmers are tearing their hair out and actually their closest relationship is with their bank manager. Well, you can say if it's true or not OK?

Emma: No, it's true, but as a farmer it's your choice whether to start a business or not isn't it?

Farmer: Yes, that's true.

Emma: So yes, you choose whether you want to get involved with the banks.

The contesting character of the student's reaction is visible in the word 'but'. By emphasising that farmers have a choice whether to go along with this tendency, Emma added something new to the discussion. The farmer's confirming move ('Yes, that's true' – see chapter 11) encouraged her to continue her argumentation and further explain what she meant. At the beginning of the tour, the students were a bit reluctant to contest the farmer's arguments. Yet, as the conversation continued and the farmer repeatedly used controversy-creating moves, students increasingly voiced divergent points of view and more and more students got involved in the argumentation. As such, controversy-creating moves can be seen as a way to create a specific, argumentative discourse where disagreement is openly expressed. If a teacher wants to orient the deliberation towards a very specific focus on the *political* dimension of ESE, s/he can do so by combining controversy-creating moves with two other kinds of political moves: 'hierarchizing moves' and 'excluding-including moves'.[1]

The following example illustrates a hierarchizing move. At the start of the guided tour, the farmer introduced the idea of what he calls 'the three Ps', which stand for 'planet', 'people' and 'profit': three concerns, he argued, that you must take into account when you work as a farmer. In his view, he said, planet should be the first concern. Thus, he set up a prioritising framework for the reasoning. He urged the students to articulate their points of view:

Farmer: Now I don't know if this ties in somehow with your vision of agriculture? . . .

[He looks around in the group. Students take notes; others look at him. Nobody answers his question.]

Farmer: Shall I answer how I think you look at this? Then you can contest me if [inaudible] . . . [laughter]

Jonas: Agriculture must be productive. So much . . . not as much as possible, it's still the intention, yes to produce food and to make sure there's enough.

Farmer: Yes, so for you the P for profit takes precedence?

Jonas: Yes. [nodding]

The farmer connected Jonas's answer that 'agriculture must be productive' to the Triple P concept as an aspect of 'profit'. By making a hierarchizing move ('so for you the P for profit takes precedence?'), he continued to enforce the prioritising reasoning and made the student take a stand ('Yes' – nodding) on which concerns take precedence and, consequently, which must give way. Doing so, he highlighted that 'profit' and 'planet' are, to a certain extent, mutually exclusive concerns. Thus, by using hierarchizing moves, a teacher can emphasise that every perspective on sustainability (e.g. sustainable agriculture) inevitably implies judgement, prioritisations or decisions of inclusion and exclusion.

An excluding-including move is another example of a political move. It is an action that we can perform as a teacher to make the students contest a proposed decision of inclusion and exclusion regarding emotionally invested commitments and concerns. For example:

Farmer: I don't receive any subsidies. And I also think that it would be very good to say that we are putting an end to them.

Emma: But *you* also don't live from [agriculture]! [original emphasis] [raises her voice]

Farmer: I *do* live from it. [original emphasis]

Emma: Oh, you said yourself that you don't pay yourself a wage! [raises her voice]

Farmer: Yes but that's different. You don't need a wage to be able to live from it. I eat from it. That's a big difference. If you think I've got 2,000 euro on my account at the end of the month. I think I've got 900 euro or something like that on my account.

Emma: Yes but food alone doesn't get you far.

Farmer: No, but yes, that's what we have to do. That's the transition we have to make. That's the change we have to bring about. I think some major steps are going to be necessary to consciously address or handle it.

Jonas: Not everyone can do it though. It is nevertheless . . .

Farmer: Why not?

Jonas: What would we eat? If everyone . . . There's more, I mean yes . . .

Farmer: Then I'd say yes, ninety percent of farming throughout the world is managed like this.

Emma: Yes, there are also I don't know how many going hungry.

With the excluding-including move, 'I don't receive any subsidies. And I also think that it would be very good to say that we are putting an end to them', the farmer suggested an exclusion of something that is highly valued by many of his colleagues: financial support via subsidies. Emma responded by contesting the suggested exclusion. It is easy to have this opinion, she argued, if one does not live from a farm, as she meant the farmer was not doing. As both the content of her intervention ('*you* don't live from it') and the raising of her voice showed, she was emotionally engaged in the debate and had no indifferent relation to what the farmer put at stake. Indeed, earlier in the conversation she told that her parents ran a conventional pig farm. Her family's livelihood depended on agricultural subsidies.

Performing excluding-including moves is thus another way for teachers to highlight the mutual exclusivity of different concerns and commitments. We can use it to make manifest the decision of which concerns and commitments to include or exclude. The dialogue above shows that both the farmer and the students were deeply engaged in this discussion. There was something at stake for them which became visible, for instance, in their strong expressive mode (e.g. 'I *do*') and the raise of the student's voice. This move also enforced the on-going argumentative conversation in which the farmer and the students regularly contest each other's standpoints. Here, they disagree about what it means to earn a living and whether the economic model suggested by the farmer could be mainstreamed. In this example, the educator creates these effects by suggesting an exclusion (of subsidies). It is easy to imagine, however, that also a suggestion of an inclusion can cause controversy over the question what to care about. For example, a proposed inclusion of the concern for dispelling hunger could be a way to enact the tension between concerns for productivity and care for the planet.

The farmer showed a strong willingness to engage in conflict-oriented deliberation and was not reluctant to voice his own, personal standpoints, opinions and passions. For example:

> That's why I also chose the farming business. Well, like, how can you change the world, eh? When you're 45, you think, right, what else is there for me to do in life and then I thought, OK, that's what I'm going to do. That's something I am really going to go for. I am really going to do it well, you know? And whatever the cost I will shoot myself in the foot and earn less. I used to work in TV where I earned about eight times as much. And now I don't, but OK, I am happier and I really feel like I am doing something useful.

In his way of arguing, he simultaneously showed a strong commitment to what he believed in *and* a consistent openness to listen to others' perspectives. As he described it himself, 'Well that's my point of view. You don't have to agree with it, do you? But that's all I can tell you'. The performed teacher moves that made it possible to stage such a conflict-oriented deliberation, are summarised in Table 13.1.

Obviously, our interventions as a teacher are not the only thing that open up or close down the space for raising and defending conflicting standpoints. Also

Table 13.1 Political moves to stage a conflict-oriented deliberation

Function	By these political moves	Description
Judging	Controversy-creating move	A teacher's action that makes the students create, express and defend conflictual standpoints
	Hierarchizing move	A teacher's action that makes the students prioritise amongst different alternatives and thus create a hierarchy of concerns by taking a stand on which concerns take precedence and which must give way
	Excluding-including move	A teacher's action that makes the students contest a proposed decision of inclusion and exclusion regarding emotionally invested attachments

other aspects have a major influence, such as the students' behaviour and the purpose of the activity. If Emma, for example, would not have responded as she did, the farmer's teacher moves would perhaps not have created this conflict-oriented deliberation. Hence, political moves should not be understood as a recipe for the right teaching technique through which conflict-oriented deliberation or political moments can and will be staged. Although as teachers we might have this intention, the outcome is always unpredictable and to a certain extent uncontrollable. Thus, it is something that cannot be completely prepared and planned in advance. Often, it is a matter of fruitfully using the students' actions as a resource for further teaching and learning. As the next example shows, we can also make other choices and thereby create very different learning opportunities, for example by performing moves that steer the activity away from conflict-oriented deliberation and from (emerging or potential) poignant experiences. This may depend on the context and purpose of the education practice.

Neutralising poignant experiences of the conflictual into normative deliberation

In this section, we discuss an example of a very different teaching practice. Here, the teacher does not perform political moves aimed at staging conflict-oriented deliberation. Instead, his actions shape the activity into an argumentative discussion oriented towards an a priori specified desired outcome, i.e. normative deliberation. By doing so, emerging experiences of the conflictual do not become manifest – and the opportunity for the students to learn from an inquiry into poignant experiences is not grasped.[2] Instead of learning from a first-hand experience of the conflictual aspects of sustainability issues and from the strong feelings involved when one is personally attached to a specific opinion, concern, etc., the learning is here focused on specific skills and knowledge. The teacher makes that possible by performing specific epistemological moves. In the conclusion of this chapter, we further elaborate how these two contrasting examples have a strong influence on how the political is experienced very differently in both educational practices.

The examples presented below come from a workshop on the 'ecological footprint' (see chapter 1) in an adult education context (Van Poeck and Östman 2018). The workshop starts with a PowerPoint presentation through which the teacher explains the concept. Subsequently, the participants are divided into small groups to discuss behaviour clues for reducing their ecological footprints. Finally, each group reports on their findings which are then discussed with the whole group. What we see happening here is that despite the participants' attempts to raise opposite and contesting views, the teacher consistently reorients the conversation in line with the workshop's strict and detailed scenario and clear purpose: teaching the participants how to reduce their ecological footprint. He does so by performing a combination of several reorienting moves and adding epistemological moves (see chapter 11: moves that reorient the learners' learning in a different direction and moves that add something to the learning environment): reinstating moves, norm-installing moves and rationalising moves.

For example, after the presentation of their small group exercise, one of the participants explicitly demanded attention for a particular concern that emerged during their discussion:

Sara: We also thought it interesting to reflect a bit. People in poor countries are on the look-out for drinking-water all the time.

Elke: Walk for miles to get water . . .

Marie: Me too, formerly [she lived in Africa in the past]. Really. It doesn't matter now. [laughter] And we actually just stand in the shower and let all that water run down to us, what those people actually could drink. Well, if you give it a thought . . .

Elke: We flush our toilet with it.

Sara: But it would be very interesting if it would become more and more widespread to use rainwater for that.

Teacher: Let's look at them, the behaviour clues. [He shows the next slide of the presentation and starts to explain the several clues and their impact on the ecological footprint.]

These participants deliberately demanded attention for this issue and they repeatedly and explicitly expressed their concern. This shows that it was not indifferent to them. Marie even voiced a very personal experience and commitment. The teacher's response to the outcome of their discussion ('it would be interesting if it became more widespread to use rainwater') was to reorient the focus towards explaining the behaviour clues in terms of facts and general guidelines that were part of his presentation and the workshop's scenario. By performing this 'reinstating move', he reinstated the participants' attention from particular, emotionally invested concerns, commitments and experiences towards the a priori specified desired outcome: knowledge on how to reduce their ecological footprints by applying specific behaviour guidelines. In the remainder of the conversation, the participants did not return to the issue of drinking water. The teacher explained the content of his presentation and the next intervention of a participant concerned a new topic, related to one of the clues addressed in the workshop's script. By performing this reinstating move, the teacher thus closed down the possibility for the participants to further articulate and explore emotionally invested concerns and commitments.

Another group reached the conclusion that their ecological footprint would benefit from using their cars less and biking or walking short distances:

Ellen: One obvious improvement would be to bike or walk short distances.

Teacher: What would you consider a short distance?

Ellen: Going to the bakery, for instance.

Teacher: How much is that in miles?

Ellen: One and a half?

Teacher: No, let me help you out: in fact, we should bike any distance under 3 miles.

Caroline: Hello-o!! [laughter]

After performing specifying moves ('What would you consider a short distance?'; 'How much is that in miles?' – see chapter 11), the teacher was not satisfied ('no') with the outcomes of it. He reacted by introducing a norm: 'we should bike any distance under 3 miles'. Thus, he indicated that this is not a matter of divergent opinions but a matter of fact ('in fact') that everyone should behave according to this standard. Such a 'norm-installing move' makes the participants react and take a stand on the postulated standard about how to behave in a certain situation. By performing it, the teacher brought in new information that rendered the conversation into a normative discourse. Caroline's reaction ('Hello-o!!') indicates that she considered the norm raised by the teacher unreasonable. Yet, instead of, for instance, urging her to clarify her disagreement, he continued as follows:

Teacher: Why 3 miles? Because cars consume most over short distances.
Ellen: Then we also had to say why we found it difficult. We found it can be time-consuming at times.
Teacher: Remember the word I just used: planning?

The teacher thus raised a fact as a backing of the norm he had suggested. Ellen responded to this claim by delivering an alternative backing for another, yet conflicting consideration: it is 'time-consuming'. Next, the teacher neglected the emerging controversy by arguing that it is all just a matter of 'planning'. By performing this 'rationalising move', he reoriented the conversation from a discussion over conflicting concerns and commitments towards a discussion about facts. His statement that it is all a matter of planning left the participants with one, normative conclusion: since it is only a matter of planning, everybody should bike under the distance of 3 miles. The teacher's moves thus reoriented the argumentation from the political register – introduced by repeated actions on the part of the participants raising alternative and conflicting concerns – towards the rational ('true or false') and the moral ('good or bad'/'right or wrong') register. Divergent legitimate concerns were thus reduced to consensus based on undisputable facts or moral principles for proper behaviour. The discussed teacher moves are summarised in Table 13.2.

Conclusion

We have presented two teaching practices in which teachers' moves very differently affect how the political is experienced in educational practice. The farmer in the first example performs political moves – controversy-creating, hierarchizing and excluding-including moves – in order to give shape to the guided tour as a conflict-oriented deliberation in which the students raise, defend and explore conflicting standpoints regarding sustainable agriculture. So doing, he stages a specific inquiry, making the students focus on the issue of sustainable agriculture as a matter of public concern in which different, conflicting concerns and commitments are entangled. It offers them the occasion to experience that there is something at stake: passions, commitments, values, interests, ideals, concerns, etc. Education, then, moves beyond cognitive reflection but also involves deeply

Table 13.2 Epistemological moves to stage a normative deliberation

Functions	By these epistemological moves	Description
Adding	Norm-installing move	A teacher's action that makes the students react and take a stand on the postulated standard about how to behave in a certain situation
Reorienting	Reinstating move	A teacher's action that makes the students reorient their attention from particular, emotionally invested concerns, commitments and experiences towards 'the lesson'
	Rationalising move	A teacher's action that makes the students take a stand concerning a factual justification for a proposed norm: accepting the justification or delivering a factual reason that justifies a divergent opinion

felt, personal experiences. The students are offered the opportunity to experience their own and other people's 'attachments' (Marres 2005); i.e. how they are affected by something by which their pleasure, fate, way of life and perhaps even the meaningfulness of their world is conditioned. By being exposed to the confrontation of mutually exclusive attachments, the students can experience antagonism while realising that one's passions and commitments run the risk of not being taken into account by others. Jointly facing situations that imply inevitable decisions of inclusion and exclusion regarding what to care about can thus give rise to a poignant experience which, as we have elaborated in chapter 10, can be a strong starting point for learning. It can instigate an 'inquiry' that makes the students reflect and actively work on their experience so as to make it intelligible.

The reinstating, norm-installing and rationalising moves performed by the teacher of the ecological footprint workshop steer learning in a very different direction. The participants' attempts to bring conflictual personal concerns and commitments into the conversation are repeatedly countered as the teacher reorients the conversation towards a normative deliberation. Thus, his teacher moves create a learning environment and stage a particular inquiry (see chapter 11) with a focus on specific knowledge, facts, clues and guidelines for reducing one's ecological footprint. As a result, expressions and poignant experiences of the conflictual aspects of sustainability issues are neutralised. They are not part of the learning environment that the teacher creates through his 'scene-setting' (adding moves: norm-installing moves) and in the staged inquiry he directs the participants' attentiveness away from them (reorienting moves: reinstating and rationalising moves). Even though the potential for learning from poignant experiences was present (instigated by the participants' interventions), the teacher's moves de-politicised this potential by systematically reorienting the conversation into the moral or rational sphere. That is, he redirects the participants' focus from a diversity of conflicting yet legitimate standpoints regarding how to organise society (the political) towards matters of 'good or bad/right or wrong' (the moral) or 'true or false' (the rational).

Different teaching practices thus influence the direction of students' learning through the specific privileging process (Wertsch 1993) that is encouraged by the teacher's moves. In the ecological footprint workshop, the teacher's moves affect the privileging process in line with the workshop's a priori defined content and purpose: learning how to reduce one's ecological footprint. Thus, what appears as reasonable, correct and relevant considerations in relation to the issue at stake here are specific knowledge, norms, facts and behaviour guidelines that are well-known by the teacher in advance. Hence, his moves serve to transfer these to the learners and to reorient the activity in the envisioned direction if necessary. As such, divergent values and concerns are often ignored. The farmer, on the other hand, influences the students' privileging process in another direction. What appears as important to take into account in his lesson is the diversity of passions, commitments, values, interests, ideals, concerns, etc. at stake, the entanglement of irreconcilable, mutually exclusive attachments and the need to make prioritisations and decisions that imply inclusion and exclusion regarding things that are highly valued. This opens up a space for poignant experiences. Students can learn from this by making meaning of their immediate experiences thereby, for example, discovering, articulating and reflecting upon their own and others' personally felt concerns and commitments.

Acknowledgements

The authors wish to thank their colleagues from the research environment SMED – Studies of Meaning-making in Educational Discourses – for their very valuable comments on an earlier version of this chapter. The research on which this chapter is based was supported by Ghent University's special research fund (research project 'Urban sustainability transitions as spaces for experiential learning' – Grant BOFSTA2016001001) and FORMAS (research project 'Wicked problems and educative spaces for urban sustainability transition' – Grant 2016–00992_3).

Notes

1 Combining controversy-creating moves with other types of interventions could privilege, for instance, an ethical or epistemological instead of political argumentative discourse in which people contest each other's position with ethical or fact-based arguments (see e.g. chapter 12: Ethical Moves).
2 This should not be considered as criticism of the teacher's practice. He may have had good reasons for not performing political moves in this particular case, for example the specific purpose of the workshop, its strict scenario, the expectations of the organisers of the activity, a lack of time, etc. Nevertheless, we discuss the example as it further illuminates the emergence of (potential) poignant experiences in the classroom and how the choice of whether to perform political moves steers the students' learning in different directions.

References

Håkansson, M. and Östman, L. (2018) "The political dimension in ESE: The construction of a political moment model for analyzing bodily anchored political emotions in teaching and learning of the political dimension". *Environmental Education Research*, pre-published online: https://doi.org/10.1080/13504622.2017.1422113

Håkansson, M., Östman, L. and Van Poeck, K. (2018) "The political tendency in environmental and sustainability education". *European Educational Research Journal*, 17(1): 91–111.

Marres, N. (2005) *No Issue, No Public: Democratic Deficits After the Displacement of Politics*. Unpublished PhD Thesis, University of Amsterdam. (https://pure.uva.nl/ws/files/3890776/38026_thesis_nm_final.pdf). Accessed 24 April 2019.

Mouffe, C. (2005) *On the Political*. Routledge, London.

Todd, S. (2010) "Living in a dissonant world: Toward an agonistic cosmopolitics for education". *Studies in Philosophy & Education*, 29(2): 213–228.

Van Poeck, K., Goeminne, G. and Vandenabeele, J. (2016) "Revisiting the democratic paradox of environmental and sustainability education: Sustainability issues as matters of concern". *Environmental Education Research*, 22(6): 1–21.

Van Poeck, K. and Östman, L. (2018) "Creating space for 'the political' in environmental and sustainability education practice: A political move analysis of educators' actions". *Environmental Education Research*, 24(9): 1406–1423.

Wertsch, J.V. (1993) *Voices of the Mind: A Sociocultural Approach to Mediated Action*. Harvard University Press, Cambridge, MA.

14 Classroom discussions

Students' learning in argumentation about ethical and political aspects of sustainability issues

Karin Rudsberg and Johan Öhman

Introduction

Many scholars have argued for pluralistic and participatory approaches to ESE (see chapter 5). The striving for more student participation has led to an increased interest in students' learning in argumentative discussions (Öhman and Öhman 2013). For us as teachers this evokes the following crucial questions: What do students actually learn through argumentative discussions? What is the function of knowledge in this kind of discussion? What is the role of peers in the individual student's learning? In this chapter we address the above questions using results from three different classroom studies of argumentative discussions about sustainability issues in upper secondary schools (Rudsberg et al. 2013; Rudsberg and Öhman 2015; Rudsberg et al. 2017).

However, before tackling these questions, we need to understand what is meant by an argument. For this purpose, the works of the British philosopher Stephen Toulmin are particularly helpful (Toulmin 1958/2003). His argument pattern (Figure 14.1) has been used in a number of classroom investigations, especially in science education (Nielsen 2013). According to Toulmin, an argument consists of several elements, each of which has a specific function. In order to be perceived as an argument, an utterance needs to include at least three elements. First, the utterance needs to contain a *claim*, which can be seen as the statement or conclusion that is argued by the student. Second, the utterance needs to include some kind of *data*, i.e. facts or other establishments about the world that supports the claim being made. Third, there needs to be a bridge between the claim and the data; a reason that establishes the connection between them. Here, *warrant* serves to legitimise the argument being put forward, i.e. the reasons that establish the connection between claim and data. According to Toulmin (1958/2003, 90), data can be identified by asking, 'What have you got to go on?' while warrants can be identified by asking 'How do you get there?'

An argument can also be more complex and include other components, such as rebuttals and qualifiers. A *rebuttal* is a counterargument to other people's arguments, or a response to a counterargument. In this way, a rebuttal functions as a safety valve in that it anticipates possible objections to an argument. A *qualifier* is part of an argument that underlines the limits of the claim, for example

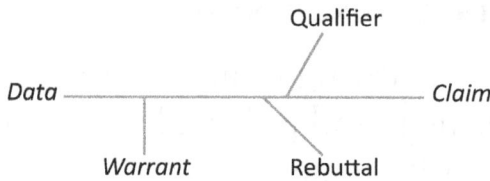

Figure 14.1 Toulmin's argument pattern
Source: (Toulmin 1958/2003)

expressions like 'in most cases', or 'tend to'. In this way, a qualifier functions as a way of regulating the certainty of the claim.

The use of rebuttals, qualifiers and data can be seen as quality indicators when evaluating students' arguments and their progress in a discussion (see Rudsberg et al. 2013). Rebuttals and qualifiers are elements that relate to the validity of an argument, its complexity and its nuances. When using rebuttals, the students show an awareness of other alternatives and possibilities (see also Erduran et al. 2004). Students can increase the credibility of their argument by adding more and improved data during the discussion.

In the following, we use this framework to discuss the content and processes of learning, the function of knowledge and the role of peers in argumentative discussions.

Students' learning in argumentative discussions

Students' learning processes in a discussion can be described as an increase in quality in terms of the *complexity, nuance* and *clarity* of the arguments that are formulated. This progression can take different forms, and in this section, we give two examples.

The first example describes a learning process in which the student holds on to a specific claim, but in response to other contributions in the discussion *gradually deepens and extends the knowledge that supports the claim.* The discussion took place in an upper secondary school, where the teacher initiated the discussion by claiming that 'improving technology is better than changing lifestyle if you want to solve environmental problems'. In the subsequent discussion, the students were expected to take a stand in relation to this statement and defend their individual points of view. One student, Kim, directly responded to the teachers' initial claim by stating:

Kim: One suggestion doesn't have to exclude the other as we just mentioned, it's not that. But then, we could be more precise and say that we have to, we have come so far now in the development, that we cannot go backwards we have to go forwards, and then this might be the only solution because we might not be able . . .

In this argument, Kim first makes it clear that there is no need to talk in terms of either/or. Instead, he advances the belief that there can be parallel solutions to a problem. This belief is related to the question about how best to solve environmental problems, which in the discussion functions as a rebuttal because it answers the potential counterargument that we have to choose between a change of lifestyle and improvements in technology. Further, in his argument Kim recalls that technology has improved so much that humans cannot go backwards, which functions as data. The claim that is established here is that the only way to solve environmental problems is to improve technology. However, this claim includes the qualifier 'might be', which limits the strength of the argument. The inclusion of these qualifiers shows an awareness that there are other alternatives and possibilities.

The teacher then asks whether Kim's statement, 'then this might be the only solution', refers to technology. Kim continues:

Kim: Yes exactly, technology. We can't change our lifestyle to the extent that we exclude cars and that so many things have been developed today that we need, sort of, that we've lived with this technology, sort of.

Here Kim claims that technology is the solution to environmental problems. From the first argument, Kim re-actualises that technology has been improved and that it is something that humans need. This functions as data. The teacher then asks whether Kim thinks that people would be prepared to change their lifestyles if environmental problems became acute. To this question, Kim responds:

Kim: Yes, to some extent I think. But I still think that we are so kind of inter-twined in this net that we can't get out of it, sure we might be able to hitch ourselves up a few notches but we'll probably never get out of it completely. We will probably never be able to go back and live like in the Middle Ages.

In this second argument, Kim makes use of prior knowledge and thereby extends the support for the claim. The question makes the student reformulate the claim and include a rebuttal, 'humans might hitch themselves up a few notches', in order to anticipate the teacher's and his peers' objections. Kim also adds support for the statement by using re-actualised knowledge in terms of an establishment about the world: the technological development society and that humans are intertwined with technology. In this way, Kim holds on to and, due to the teacher's question, reinforces his initial standpoint. This is an example of how an argument can improve during discussion as a result of developing and adding knowledge that supports the claim.

In the second example (transcripts collected from Rudsberg et al. 2013) another student, Marcus, also *adds new data* to his argument, although in this case his argument takes a new direction when he *suggests new solutions* by widening the teacher's initial claim and adding explanations that limit the claim's application. In his first utterance, Marcus states:

Marcus: If it is possible to solve all environmental problems with better technology then we should of course do that.

We can see that this first statement is not supported by any data and is more rhetorical in nature, in that most people would agree with the claim. The statement is also in line with the teacher's first dichotomised claim. In his second utterance, Marcus redefines the first statement by adding uncertainty about putting 'trust in scientists coming up with better methods', but then argues that, 'it is better than just lowering our standards'. However, in his third and fourth utterance the argument is developed by the inclusion of a third alternative, legislation:

Marcus: Well, a thought just struck me, now I am not responding to any problem or any kind of argument, but . . . If you introduce new laws that say that you can only shower for ten minutes per day, that kind of in some way changes our lifestyle, or so that these environmental problems are solved. How, how, I think, isn't this better than, you still work with technology and a change . . . [hesitates and doesn't complete the sentence]
Teacher: You mean a third way, to use legislation to . . . [does not complete the sentence].
Marcus: I don't know, technology is progressing anyway and research is going on all the time. Now I was thinking that maybe we should give more money to solve these problems, but yes if we made laws that would make everybody sort of . . . [does not complete the sentence].

In the third argument, Marcus introduces a new option by re-actualising the earlier knowledge about legislation. In this way, he finds new solutions by exceeding the original dichotomised statement. In the progress of the argumentation Marcus adds more elements, namely the qualifiers 'kind of in some way' and 'still work with technology', in order to validate the argument being made. Finally, in the fourth argument the claim is developed further when Marcus states that more money should be invested in technological research. The argument is here specified by the use of the qualifiers 'maybe' and 'I don't know', thus clarifying the limits of both the data and the claim.

The above two examples show how students' arguments develop during a classroom discussion in different ways. During the discussion the students gradually make their arguments (i) clearer by specifying the boundaries of their claim, (ii) more nuanced by adding, deepening and extending the knowledge that supports their claim and (iii) more complex by putting the knowledge they use into a wider context. In this learning process, knowledge is both of crucial importance and has direct relevance for the quality of the argument. In the next section, we take a closer look at the role that knowledge plays in argumentative discussions and for how students formulate their arguments.

The functions of knowledge in students' argumentative discussions

As stated above, data is a necessary element in an argument and, in this context, knowledge has the general function of justifying a claim. However, when analysing the function of knowledge in an argumentative discussion in more depth, we found that knowledge has at least *six specific functions* in the argumentative process (Rudsberg and Öhman 2015). These functions influence both the direction and depth of the collective argumentation. In the following we describe and exemplify these specific functions by using transcripts of students discussing climate change and the difficulties of international climate change agreements (transcripts collected from Rudsberg and Öhman 2015).

First, knowledge can contribute to *emphasising the complexity* of the discussed topic:

Maria: Yes, exactly. [inaudible] in the end I really do think that there is every reason to succeed, although not perhaps from our ideological standpoint, if I can call it that. At the same time it is so very difficult because everything is connected, so if we do something there will be consequences somewhere else. [. . .] But perhaps that's something we just have to accept, because the consequences that follow on from this are part of something much bigger, and are something that we can compensate for. But I also think that it's quite difficult, because we've seen how we tackled industrialisation, how we developed, we have never seen any other development happen in such a way. This is why it is so difficult to find the ultimate leapfrogging, or whatever it's called, that won't have any consequences at all. Since we know so much, we also have a lot to take into account; social prosperity, material wealth, geographical things, our environment, economic growth, that is to say that individuals should be properly catered for and have political freedom; we need to take everything into consideration. The question is whether this is possible.

Here, Maria's recalled historical and political knowledge serves to emphasise the difficulties of finding solutions to climate change issues due to their complexity and the multitude of environmental, social and economic aspects that need to be taken into consideration at both an individual and societal level.

A second function of knowledge is to *highlight conflicting interests*:

Thomas: I think that there are two things to flag up: First, you talked about animals, what is actually more important? And today I read in *DN* [a newspaper], it was something, just a tiny thing that happened in

Africa in a country where it was . . . Anyway it was about organising it again so that you can shoot animals to a large extent because, that the farmers, they're suffering from these animals, elephants, coming in and trampling on things. You can think like this, should I think about the animals then, I mean in the general ecosystem – or is it better for farmers to have better economic conditions? That's what we have to weigh up . . . For us it's perhaps more . . . A person can think that the ecosystem is more important and they can say 'but our life is more important' and that it's about development in general, everyone wants a respectable life.

By drawing on knowledge in a recent newspaper article, Thomas highlights two conflicting interests: preserving biological diversity and the living conditions of African farmers. Furthermore, Thomas argues that this conflict would play out differently if you lived in a wealthier part of the world or in poorer conditions.

A third function of knowledge is to *provide evidence in a counterargument*. In this example, one student has claimed that oil-producing countries need to convert from oil to other sources of income. This is exemplified by the way that Dubai has invested in tourism and climate-smart cities. In response, Kate argues that tourism is not always environmentally friendly:

Kate: That's not so environmentally friendly either. In the first place you have to fly there and then you can't live in a city in a sustainable way. [Inaudible] but in the desert it's even harder, because then you have to pump groundwater all the time.

In this case, Kate's environmental and human geographical knowledge enables her to formulate a counterargument to the claim that tourism is an unsustainable substitute for the oil industry in the Middle East.

In addition to the functions exemplified above, knowledge can also serve to *clarify and correct* earlier statements, *predict consequences* of a claim and *add support to an earlier claim* (see further Rudsberg and Öhman 2015).

From these examples, and those in the previous section, we can conclude that two kinds of simultaneous learning processes take place in argumentative discussions: *learning to formulate an argument* and the *learning of knowledge content*. In these kinds of discussions, the students do not only reproduce knowledge, but also use it to position themselves argumentatively in relation to crucial human and environmental problems. More specifically, by participating in discussions students can learn how to *use their knowledge in a deliberative practice*, connect their *knowledge to value judgements (claims)* in an argument and *contextualise their knowledge* in relation to an important question. This means that their knowledge can receive an extended and deeper meaning. In sustainability discussions (deliberations), students get the opportunity to use knowledge that they have learned in other lessons, for example knowledge about politics, environmental studies, history, biology and human geography. In the students' arguments, knowledge from these different areas are interwoven into a complex reasoning in an argumentative context. By being critical of each other's arguments, the students also increase their awareness of the trustworthiness of different sources of information (such as social media and the internet in general). However, it can also be the case that the data a student uses to support his or her claim is incorrect or is based on preconceptions or pure guesses, and that this goes unnoticed by peers. Here it seems reasonable for the teacher to intervene in the discussion and question and correct such inaccuracies.

So far, we have mostly been concerned with individual students' learning and use of knowledge in argumentative discussions. However, in these discussions, learning takes place in the encounters within the social environment and foremost with peers. In the last section, we focus on the social aspect of deliberations and especially the importance of peer interactions.

Peer interactions in argumentative discussions

Interactions between peers are important for the individual student's learning because it is in relation to other students' arguments that knowledge is re-actualised and own arguments are developed (Rudsberg et al. 2017). In many cases, the students actively elaborate on other people's reasoning. In this way, students develop their arguments by for example *paraphrasing a shared position, completing the arguments of others, elaborating on others' views, showing that the opinions of others have not been satisfactorily justified* and *critiquing other students' reasoning for missing an important distinction.*

This social aspect of learning in and through argumentative discussions is illustrated below by examples collected from the same student discussion on climate change and the difficulties of international climate change agreements as used above (transcripts here from Rudsberg et al. 2017). The first example illustrates how Tom *criticises other students' reasoning for missing an important distinction* and uses this as a way of strengthening his own standpoint:

Tom: But it's really important to put aside what's done, I mean you must bear it in mind as well, but what has happened has happened and what we have to do now is to make a big difference. I mean absolutely that, if you think that we have an historic debt in pollution, and that's something that is true and we really have to do that. But we anyway have to have everyone on board in some way. And even if we have had enormous historical debts and we had been as poor as all the other countries, then I don't know whether I'd have thought that we should do so much more anyway. But now it happens that we can do so much more because we have so much money then it's obvious. But to keep on looking at what has happened and relations and all that blah blah blah, that's probably not what is important now . . .

Here Tom inserts a distinction to knowledge used by another student concerning responsibility for solving environmental problems. This is done by adding a rebuttal about the unimportance of historical guilt. Tom then adds a new statement about the importance of looking at the present situation concerning climate change supported by re-actualised knowledge about prosperity as a reason for doing more about it.

In the second example, Sandra contributes to the subsequent discussion by *elaborating on others' views*, in this case the unimportance of historical guilt as stated by Tom, above:

Sandra: Well, yes, but I think that you can't really disregard it. You must, well, it would be unfair to say 'yes, but, sorry, okay now we have a problem, you'll just have to forget that we've done a load of idiotic things, now you'll also have to reduce your emissions enormously but we, we would like to buy up so that we can continue to pollute the atmosphere and benefit from you reducing yours through different projects'.

In response to this, Jane *completes Sandra's argument* that it is unfair to expect developing countries to reduce their carbon dioxide emissions:

Jane: Mmn, absolutely, but it still needs everyone to do something, that everyone . . . I mean it's all about sacrifices as well, that you are prepared to sacrifice certain things, and in a certain way perhaps . . . Undeveloped countries haven't come as far in their development, perhaps haven't contributed as much historically, but . . . And sure, that some perhaps have a greater responsibility because they can do more and therefore ought to do more, but I mean it's still works so that . . .

These examples illustrate how students can create an argumentative environment in which they together formulate complex and nuanced lines of reasoning that contain interrelations between, for example, history, politics and the economy. In this particular discussion, the students' different actions guide the subsequent arguments in the discussion and contribute to multiple ways of dealing with the problems and agreements around climate change. This argumentation is characterised by seeing climate change issues as involving different value perspectives and ideas. The students talk about the difficulties of international climate change agreements as a dilemma that has no obvious solution. In their argumentation, the students focus on the relation between political and economic aspects and how these are connected at both a global and a local level. Through the elaboration of these relations, the students together create a complex understanding about climate change issues (see further Rudsberg et al. 2017). However, there is also a risk that the students' strivings for consensus will reduce the diversity of the discussion (Öhman and Öhman 2013). It is therefore important that the teacher actively challenges the common view and allows for possible alternatives. The role of the teacher can thus be seen as guiding students' discussions to a point where differences are displayed by encouraging them to express their personal opinions, confront each other, be critical and discuss conflicting ideas (see also chapters 11, 12 and 13 for further elaborations on how this can be done).

To conclude, the learning that takes place in argumentative discussions is a collective and communicative process. The interactional aspect of individual student's learning becomes visible in the way a student includes parts of other students' arguments in his/her own and responds to other people's arguments when formulating counterarguments. The individual student's contributions to the collective discussion becomes visible when peers connect to and take the student's reasoning further in the subsequent discussion.

Concluding remarks

In the introduction to this chapter, we asked three crucial didactic questions about argumentative discussions: What do students actually learn through argumentative discussions? What is the function of knowledge in this kind of discussion? What is the role of peers in the individual student's learning? Based on recent research studies and using examples from real classroom discussions, we have shown:

- that students can learn to *formulate an argument* and gradually make it *clearer, deeper and more complex*
- that knowledge plays a crucial role in these discussions and that students can learn how to *use their knowledge in social practice*, connect their *knowledge to value judgements* in an argument and *contextualise their knowledge* relation to an important question
- that students' *learning progress is strongly connected to the interaction with peers* when they respond to, develop and criticise other students' arguments in their own argumentation.

Taken together, we can conclude that argumentative discussions can be seen as an important way of facilitating learning about complex sustainability issues with ethical and political implications. We have also argued that students do not only learn content knowledge about these issues, but also how to take a stand and formulate valid arguments, thereby equipping them to participate in deliberations about crucial questions in the future. In this way, participation in teacher-led argumentative discussion can be seen as a way of developing students' democratic action competences.

Insights into the learning processes that take place in argumentative discussions are essential for teachers working with a pluralistic and participatory approach to ESE (see chapter 5). By actively participating in a discussion with different ethical and political moves, the teacher can deepen and nuance the students' standpoints, challenge the common view and introduce alternative possibilities and opinions (Rudsberg and Öhman 2010; Öhman and Öhman 2013; see also chapters 11, 12 and 13). In this way, the teacher plays a vital role in the *quality* and *diversity* of the discussion and also in students' growth as democratic citizens: being knowledgeable as well as skilled in deliberations.

References

Erduran, S., Simon, S. and Osborne, J. (2004) "Tapping into argumentation: Developments in the application of Toulmin's argument pattern for studying science discourse". *Science Education*, 88(6): 915–933.

Nielsen, J.A. (2013) "Dialectical features of students' argumentation: A critical review of argumentation studies in science education". *Research in Science Education*, 43(1): 371–393.

Öhman, J. and Öhman, M. (2013) "Participatory approach in practice: An analysis of student discussions about climate change". *Environmental Education Research*, 19(3): 324–341.

Rudsberg, K. and Öhman, J. (2010) "Pluralism in practice. Experiences from Swedish evaluation, school development and research". *Environmental Education Research*, 16(1): 95–111.

Rudsberg, K. and Öhman, J. (2015) "The role of knowledge in participatory and pluralistic approaches to ESE". *Environmental Education Research*, 21(7): 955–974.

Rudsberg, K., Öhman, J. and Östman, L. (2013) "Analysing students' learning in classroom discussions about socio-scientific issues". *Science Education*, 97(4): 594–620.

Rudsberg, K., Östman, L. and Aaro-Östman, E. (2017) "Students' meaning making in classroom discussions: The importance of peers". *Cultural Studies of Science Education*, 12(3): 709–738.

Toulmin, S.E. (1958/2003) *The Uses of Argument*. Cambridge University Press, New York.

15 Power and governance in environmental and sustainability education practice

Marie Öhman and Johan Öhman

Introduction

Chapter 5 describes three selective traditions within environmental and sustainability education: a *fact-based* tradition, a *normative* tradition and a *pluralistic* tradition. The traditions represent different views of environmental and sustainability problems. In the *fact-based* tradition, teachers primarily treat environmental issues as knowledge problems that are caused by a lack of scientific knowledge about how nature works and the impact that modern society has on ecosystems. The *normative* tradition maintains that environmental problems are largely due to the lifestyles that have developed in industrialised and post-industrialised countries. In the *pluralistic* tradition, environmental and sustainability problems are mainly seen as political conflicts between different knowledge perspectives, values and interests.

Knowledge plays an important role in all three traditions, although they all have different practices relating to the role of (scientific) knowledge, for example as a solid foundation for informed decisions and actions, as motivators and arguments for a sustainable lifestyle, or as sources for critical discussions about different kinds of societal change (see chapter 14).

Despite the differences, in all three traditions teachers have to decide which knowledge content to present to their students, how this should be done and how to deal with the different values and norms that are included in the traditions.

Of necessity, teachers have to think about the choice of knowledge content in their teaching. The choice of content is of course related to the syllabus, but although syllabi provide guidance, the content itself has to be selected by the educators themselves. These choices are also made in relation to the teaching habits and selective traditions that are followed and are influenced by personal experience and the students' own interests (see chapter 5). In this way, some subject content is prioritised and regarded as more natural and obvious than others.

The choice of knowledge content is never neutral or value-free. Knowledge is created from certain perspectives, ideas and interests and is thereby always linked to norms and values that in one way or another indicate what is important, unimportant, good, bad, desirable or undesirable (see also companion meanings in chapter 4).

Thus, the argument not to include sustainability issues in the curriculum because they are so value-laden is not valid, since all education is normative and value-laden. Knowledge content includes norms and values, guides and steers students in certain directions and thereby favours certain ways of thinking and acting. This in turn facilitates or limits students' understanding of sustainable development and how to look at themselves and their social and natural environments. We can therefore say that the teaching content in school and the specific knowledge that is taught have an inbuilt aspect of *power*.

In this chapter, we elaborate on the power aspect of the knowledge content that is highlighted in environmental and sustainability education. This is done with the aid of a theoretical perspective that is mainly inspired by the work of the French social theorist and idea historian Michel Foucault (Foucault 1976/1980, 1978/1991, 1982/2002). The ideas in Foucault's work are of course much more complex than we are able to capture in the chapter. In order to make these ideas more concise and comprehensible, we have distilled and adapted them to an educational context. However, by highlighting a power perspective, the aim is to offer teachers a way of reflecting on their choice of content, the consequences of their choices and what kind of alternatives are available.

Understanding the concept of power and discourse

Power is a concept that is used in various ways in different contexts. In everyday language, the word power is often associated with something that limits people's freedom through force and oppression, and we sometimes discuss who has power, gains power, holds power or exercises power. In school, we come across discussions about whether teachers have power over students, or whether students have power over teachers. Does the principal have the greatest power, or is it the caretaker? In these contexts, power is regarded as something that is exercised from specific social positions. Important research emanating from this power perspective aims to highlight oppressive power and make the exercise of power by individuals or groups visible.

In order to understand the kind of power that is presented in this chapter, we need to jettison ideas about power that are associated with authorities, laws and the state's judicial sphere. Instead, interest is directed towards power in terms of the inclusion of a knowledge content in education and how this will affect students' thinking and acting.

Power is connected to the knowledge and language that we use in our practical, everyday activities. A frequent and dominant way of talking about the world or an aspect of the world (for example sustainable development) is called a *discourse* (MacLure 2003). One way of understanding discourses is to view them as regular patterns of speaking or acting in different situations. The concept of discourse is closely connected to specific knowledge and, thereby, a specific truth. A certain understanding of the world, an issue or a problem is taken for granted. Embracing a specific truth or knowledge about the world means being able to talk about and understand the world in a particular way.

It is through language, or the discourse, that we can understand, think about and evaluate our surroundings and ourselves. In other words, when dominating knowledge, norms and values are expressed in our actions and language use, we call it a discourse. Discourses are created because we use language regularly and act in certain ways. Our actions and language use can also reproduce the discourses and enable their power to be expressed. Different ways of talking about things result in different prerequisites for understanding, evaluating and acting. The language of a discourse 'does' something to the world, i.e. it conjures up or constructs our reality.

In our daily lives, both in and out of school, we encounter several discourses. These sometimes co-exist, although in general one of the discourses dominates. However, for the individual, there is always the possibility of opposing a dominating discourse by referring to alternative discourses. For example, the various ways in which sustainable development is talked about in the three selective traditions presented in chapter 5 can be understood as different discourses that establish different possibilities and limitations for what is imaginable when it comes to sustainable development.

The offer of a certain kind of knowledge means that certain actions will be perceived as more reasonable than others, be more likely to happen, or be more likely to be taken seriously. Power, in terms of a possible action spectrum, guides, generates or governs other actions. The power/knowledge relation is concretely expressed through its governing function – a relation between power, knowledge and the self. Or, as Foucault (1982/2002, 341) expresses it, 'To govern is to structure the possible field of action of others'. Foucault's central concept of power/knowledge should accordingly be read as one word and seen as intimately interwoven.

A specific and systematic use of language about sustainability issues thus has a governing function – it guides students in a certain direction. The offered knowledge contains messages about what sustainability is or ought to be, or what is sustainable or not. It also indicates which aspects of sustainable development – environmental, social or economic – ought to be foregrounded. In other words, the use of language and the choice of content facilitate *a* specific way of understanding sustainable development. The foundations for future actions are laid in terms of actions that affect students' possibilities to think about themselves and sustainable development. This happens through the process of *self-governance*. Once the student has accepted and acquired a certain kind of knowledge, certain actions will appear to be more reasonable, relevant, effective and justifiable than others. The external governing of the discourse by the social environment, e.g. teachers and peers, thus turns into an internal governing of the self.

Governance in ESE teaching practice

Relating power to the discourses that are expressed through language and other actions means that power is first of all dispersed – it is wherever people act. Secondly, it is understood from the point of view of how it is concretely manifested in people's use of language and other actions in a certain context. In an educational context, it means a focus on how the teaching practice guides the students' thinking in a way that facilitates or limits their present or future actions. In the process of governance and self-governance, it means that certain actions and ways of thinking become more possible than others.

Let us imagine how teachers rooted in the three different teaching traditions might start a lesson on climate change:

Fact-based teacher:	'Today we will study the heat-trapping nature of carbon dioxide and other gases and their ability to affect the transfer of infrared energy through the atmosphere.'
Normative teacher:	'Today we will learn more about how we can reduce our effect on the climate in our everyday practices.'
Pluralistic teacher:	'Today we will discuss different standpoints in the debate on climate change and scrutinise their different evidence, interests and arguments.'

If these different uses of language recur, they will call for certain responsive actions from the students. Some actions will thus be perceived as more reasonable and rational than others. In the first case, it is likely that the students would act in ways that deepen their scientific knowledge about climate change, for example by reading books and articles about the subject and reflecting in a systematic and logical way. In the second case, they may take a certain moral responsibility and gradually adopt a specific climate-friendly attitude and behaviour. In the third case, they might reflect on different alternatives and develop solid arguments for their own actions and views of climate change.

Looking at these examples from a power perspective means that a certain view of climate change is included at the same time as others are excluded. The different views also make some thinking and acting more likely than others. Power is thus about how certain actions facilitate other actions. It means that an action makes certain ensuing actions more reasonable and possible.

Power – possibilities and limitations in education

It is important to highlight that in Foucault's view power is both productive and repressive. When a teacher chooses a certain content or perspective, some things are included and others are excluded. Certain kinds of knowledge will thus become more accessible to students than others. That which is included will naturally create a number of *possibilities* for students to think about and understand sustainable development. However, certain things are also excluded, which *limits* the possibility to understand other ways of thinking about and understanding sustainable development. Content in terms of knowledge includes ideas about what kind of learning is important or unimportant, what is essential and what is not. Some ways of thinking are more privileged than others. What is important to consider in this context is that discourses and the power/knowledge relation produce a reality and a certain way of thinking and acting. Some versions of the word take precedence over others and indicate how students can relate to themselves and their surroundings. In other words, knowledge creates possibilities. However, when certain knowledge appears as given, it means that other knowledge is ruled out. What happens in a teaching practice is that the prioritisation of some knowledge results in the exclusion of other knowledge and ideas. This is the inevitable course when we as teachers have to select a content. Important, however, is to be aware of the consequences of the choice of content for the students' abilities to think about and understand sustainable development and their role in it.

For example, if ESE studies of the natural environmental are dominated by the discourse of natural sciences, a specific view of nature will be promoted. Here, nature is treated as an object that is separated from the human observer, is universal and regular and something that functions in terms of causes and effects that can be measured. Generally, these studies involve indirect encounters with nature, i.e. an encounter that is mediated through lectures, books, pictures and films. In scientific studies many things are made possible, in that they promote an understanding that is based on established and reliable facts and models. In other words, they offer a standardised and systematic way of thinking about nature and a consistent and universal language for reflection and communication.

But some things are excluded in this discourse, such as an understanding that is based on personal experience through a direct encounter with the natural environment in outdoor recreation/life/studies. Learning about nature through a direct encounter involves our bodies, senses and emotions. This understanding is relational, in that it is not only an experience of nature, but also of ourselves. In this way, nature is not reduced to facts and models, but can have aesthetic, moral, existential or emotional implications. Such an understanding can open up the human-nature dualism and make a relational ethics possible, where people's personal experiences lay the foundation for ethical reflection and a care for nature.

What we want to highlight with this example is the importance of reflecting on the dominant discourses in education/ESE and identifying what they include and make possible and what they exclude. In relation to exclusion, we also need to think about which discourses might complement the dominant discourse in order to make the students' understanding richer, broader and more nuanced.

We have now discussed how discourses (language use and other actions) that are prioritised in the teaching practice create a power/knowledge relation that both facilitates and limits students' ways of acting and thinking. The question is – what will the consequences of this be for students?

Offering students subject positions

We have already indicated how power manifests itself in people's everyday actions, and especially how these everyday actions shape people's ways of acting, thinking and being. The possibilities of power can be said to generate, promote, benefit and support people's further actions (Edwards 2008). According to this perspective, people are regarded as constituted in action, i.e. that the student 'is or becomes someone' in relation to the teaching practice in which certain knowledge is made available (see person-formation in chapter 4). Individuals are constituted in relation to certain knowledge that creates a possible action spectrum, i.e. *subject positions* are created from the narratives that are available (Wetherell and Potter 1992). In relation to the knowledge content, there is a certain amount of room for students to think and act.

In general, we can say that discourses, power/knowledge relations and governing processes are about the politics of the subject, i.e. how the individual, or the student, is governed and created in a certain context. The use of language, the knowledge content and the norms and values that are favoured in ESE guide how students ought to think and act. When students are offered a certain version of the world, it can be said that a certain character or identity is constituted (created). This means that students 'become someone' in relation to the knowledge perspective that is available to them in that specific practice. It is always in relation to how something is talked about or described that students can reflect on their lives and, so to speak, become someone.

This may sound rather deterministic – that power and dominating discourses determine people's actions. But individuals can be regarded as an effect of power at the same time as they set power in motion through their actions. In line with Foucault, we claim that people's actions are a result of certain knowledge, at the same time as people's actions shape knowledge. That means that knowledge is constituted in action and simultaneously creates predispositions for further action. This is important insofar as it dissolves a dualistic view of knowledge and people's actions and instead portrays them as mutually defined and interdependent. By viewing Foucault's power perspective in this way, we can avoid both a deterministic and autonomic view of people's actions. Things can always be changed through our language and our actions, but this must always be done in relation to what is generally most accepted. There is therefore always room for counter-power – students often have access to other discourses outside the confines of school. Students can resist the power that works through dominating discourses and thus also affect the knowledge content of the teaching.

Let us return to the three selective traditions and the different subject positions they offer, i.e. students' roles as citizens in relation to sustainable development. Within the framework of the fact-based tradition, students can imagine themselves as *informed citizens*. By emphasising the importance of scientific knowledge, normative action patterns are created that allow students to understand themselves and be understood by others as competent, sensible, rational, clever and reliable. In the same way, the normative tradition offers the subject position of *moral citizens* who are active, caring, committed and responsible. The pluralistic tradition offers the subject position of *political citizens* who are participatory, social, aware, conscious, concerned and critical.

However, the implicit opposites of these positions are also possible subject positions. In the creation of informed, moral and political citizens, *ignorant, immoral* and *careless* citizens can also be constituted. Undesirable actions are thus defined in relation to actions highlighted as the right ones. When we categorise the desirable, we also categorise the deviant. Students who are willing, have the right attitude and want to take responsibility are often the accepted ones. Students who are unwilling, indifferent and do not care are not always accepted and are often judged. However, it is important for teachers to be aware that students have different abilities to live up to the right subject position due to their will and opinion, and also their diverse cultural, social and economic backgrounds. Thus, the knowledge content that is made accessible in ESE offers certain subject positions and thereby potentially the inclusion and exclusion of students.

The power perspective as a tool for teachers' reflections

In this concluding section, we discuss how the power perspective can be helpful and used by teachers as a way of reflecting on and critically evaluating their teaching activities. We have already indicated that the teaching practice of ESE is often deeply rooted in habits, traditions and customs, and that we often regard the content as natural and obvious. With the aid of Foucault's power perspective, it is possible to reflect on and critically analyse what the activity offers, particularly that which is often regarded as obvious and taken for granted. In general, it is about adopting an approach that means that teachers can observe how students are invited to look at the world, sustainable development, themselves and other people in certain determined ways.

We believe that it is important for us as teachers to understand how dominating discourses in society or in specific groups/communities impact the discursive practices in school. A striking example is the discourse of *ecological modernisation*, which builds on a strong belief in market economy and technological innovations and proposes, 'economic growth and the resolution of ecological problems can, in principle, be reconciled' (Hajer 1996, 248). As a consequence, education becomes a political tool in social engineering for sustainable development. When talking about discourses as language rules that allow certain statements and actions to be made, teachers need to understand how dominant discourses in society (for example ecological modernisation) impact the ESE practice in terms of facilitating or limiting actions. Thus, the power perspective can be seen as a lens through which to explore *how* dominant discourses in society are produced and reproduced in education.

When teachers talk about and describe a phenomenon, one of several possible versions can be used. Here, important questions to ask are (i) In how many different ways do I talk about and describe sustainable development? (ii) What will the consequences be for the students in terms of understanding sustainable issues and their role in sustainable development? and (iii) What other possible versions and descriptions are there? In other words, it is partly about reflecting on what the specific content does with the students, and partly about thinking about what could be done differently. The power perspective thereby facilitates a practical critique that can lead to reflections about the possible consequences, where teachers can ask questions about whether it would be possible or even desirable to talk and act in other ways.

It is important that the power perspective and its critical approach are not regarded as criticism of individual teachers. We all take part in available and dominant discourses and often act in accordance with the established interpretations and 'truths' that they contain. The point is to highlight the obvious and taken for granted actions that we all help to create. It is not a question of what is bad or good, but of which knowledge is presented in the teaching and what the consequences of this might be. Further, assuming this critical position does not mean making universal claims about what is true, bad or good, but is rather an approach that offers a critical position without saying how things ought to be by coming up with better and truer solutions. Foucault also points to the difficulty of formulating a universal theory about what is true. However, this should not be perceived as a relativist perspective. Based on a specific purpose, and in a certain context, we can of course express what is good, positive, justifiable, rational and sensible. Foucault emphasises that:

> My point is not that everything is bad, but that everything is dangerous, which is not exactly the same as bad. If everything is dangerous, then we always have something to do.
>
> (Foucault 1982/1991, 343)

For Foucault, the important question is not if things are good or bad. The important thing is 'the dangerous bit', because how we speak and act leads to different consequences. By dangerous, Foucault means that some approaches may appear to be obvious and natural and are taken for granted in a way that makes us blind to other perspectives and possibilities. It is a question of challenge the privileging of certain knowledge, norms and values. This critical projection means that we lean towards that which tends to make things immobile; that which is presented to us as unassailable and timeless. In other words, Foucault's message is about challenging the force of habit and asking questions about why things are like they are and not otherwise. This is particularly important in educational situations and in the context of current 'unsustainability', where people are steered and guided by being offered certain versions of the world. A critical perspective, on the other hand, opens up the possibilities of choice for teachers, and thereby also for students. The power perspective can allow for critical reflections of what is included and excluded in our everyday teaching practices – insights that can challenge our ways of thinking and acting in relation to the environment and sustainable development.

References

Edwards, R. (2008) "Actively seeking subjects?" In Fejes, A. and Nicoll, K. eds., *Foucault and Lifelong Learning: Governing the Subject*. Routledge, London, 21–33.

Foucault, M. (1976/1980) "Two lectures". In Gordon, C. ed., *Power/Knowledge: Selected Interviews & Other Writings 1972–1977*. Pantheon, New York, 78–108.

Foucault, M. (1978/1991) "Governmentality". In Burchell, G. Gordon, C. and Miller, P. eds., *The Foucault Effect: Studies in Governmentality*. The University of Chicago Press, Chicago, 87–104.

Foucault, M. (1982/1991) "On the genealogy of ethics. An overview of work in progress". In Rabinow, P. ed., *The Foucault Reader*. Penguin Books, London, 340–372.

Foucault, M. (1982/2002) "The subject and power". In Faubion, J.D. ed., *Essential Works of Foucault 1954–1984, Volume 3, Power*. Penguin Books, London, 326–348.

Hajer, M.A. (1996) "Ecological modernisation as cultural politics". In Lash, S. Szerszynski, B. and Wynne, B. eds., *Risk, Environment and Modernity: Towards a New Ecology*. Sage, London, 246–268.

MacLure, M. (2003) *Discourse in Educational and Social Research*. Open University Press, Buckingham.

Wetherell, M. and Potter, J. (1992) *Mapping the Language of Racism: Discourse Theory and Practice*. London, Sage.

16 Teaching as a matter of staging encounters with literary texts in environmental and sustainability education

Petra Hansson

Introduction

The 2030 Agenda for Sustainable Development stresses pedagogies that empower learners to contribute to sustainable development (UNESCO 2017). This chapter proposes four pedagogical principles that aim to support teachers to organise reading and writing activities that empower students to participate in discussions of sustainability issues in environmental and sustainability education (ESE). Together, these principles provide a framework that can be used as a source of inspiration for putting reading and writing into play with students' experiences when teaching sustainability issues through encounters with literary texts. The principles build on the idea of creating a stage where students are challenged and encouraged to express opinions, emotions and values and react to and act on sustainability issues in a systematic way. The pedagogical principles advanced as a consequence of the following teaching experience:

> With a straightforward poem and three formulated reading and writing activities I entered the classroom. The first reading activity was to read through the poem and while reading choose passages the students for some reason liked and write personal reactions in their reading journals. I was not suspecting this to be difficult and viewed the activity more like a warm-up to be followed by a text analysis of the poem and finalised by composing a short poem. However, the warm-up activity turned out to be challenging. The students raised their hands one by one asking: Which passages are the correct ones to choose? I altered the course of the lesson and let the students just choose single words from the poem and write them down in their reading journals without writing reactions. However, the students were still puzzled, choosing words with great resistance and asking if they managed to choose the correct words. This took the whole lesson and we did not even finish the warm-up. What struck me the most were the students' insecurity of making choices and their fear of making mistakes and my didactical failure of assuming that the students were familiar with the suggested teaching approach and my lack of language and theories to explain the purpose of the activity. Obviously, I needed to re-think the didactical approach and develop appropriate

strategies that would help me to teach poetry in a way that could empower students to make choices and express reactions rather than puzzle them.

Since that day, the pedagogical principles to the teaching of literature presented in this chapter have been tried out and refined in language teaching, teacher education, museum education and interdisciplinary higher education, and their theoretical underpinnings have been explored in previous environmental and sustainability education (ESE) research (Hansson 2010, 2014). The chapter first provides an account of the theoretical inspirations underpinning the pedagogical principles, including ecocriticism (Hansson 2009; Garrard 2010), reader-response theory (Rosenblatt 1985, 1994, 1995, 2005) and theories of writing and responding (Elbow 1989; Dysthe et al. 2011; Hansson 2010; Manchón 2011). Second, it provides teachers with a practical description of the pedagogical principles using Daniel Defoe's novel *Robinson Crusoe* as an example of how sustainability issues can be addressed and how sustainability competences can be trained through the teaching of literature. Not only being a classical adventure story, *Robinson Crusoe* also evokes sustainability themes such as the vulnerability of small islands (IPCC 2014, 2018), the wide-ranging concern of political and ethical dimensions of human-nature relationships in times of increasing human-induced pressure on planetary boundaries as well as emotional dimensions of sustainability issues such as fear and hope.

Ecocriticism

Advancing in the 1990s, ecocriticism is a relatively new interdisciplinary field in literary and cultural studies (Glotfelty 1996). According to Ursula K. Heise (2006, 506), ecocritical inquiry encompasses three spaces of study, (1) scientific analysis of nature, (2) analysis of cultural representations and (3) the political fight for more sustainable dwellings in the world. The potentiality of literature to evoke emotions and aesthetic experiences (Dunlop 2008) and to imagine alternative ways of relating to environment and place (Wason and Ellam 2010) is viable in discussions of ecocritical pedagogy.

Early ecocriticism emphasised readings and analyses of texts portraying close human-nature connections, such as Henry David Thoreau's *Walden*, implying that values of connectedness with and appreciation of nature are vital for fostering sustainable development (Buell 1995; Garrard 2010). Even though ecocriticism has clear pedagogical ambitions to foster sustainable development, empirical investigations of students' learning in relation to ecocritical readings have not been a priority (Garrard 2012; Hansson 2009). A study that explores the process of environmental meaning-making and evocation of a sense of place in teacher students' reading of Thoreau's *Walden* shows that introducing literature in environmental and sustainability education is a complex endeavour (Hansson 2014). For example, the students' meaning-making is to a large extent produced in relation to the staging of the reading activity that points to the importance of taking both institutional, social and individual aspects into account when staging literary encounters in ESE. For ecocritical inquiry to be an incentive for sustainability transformations, this chapter argues that teachers need strategies for staging literary encounters that encourage students to cross borders between literary texts and real-world sustainability issues. In the following, the approaches to reading and writing underpinning the pedagogical principles are discussed.

Reading

Rosenblatt (1995) discusses the teaching of literature as a way to create pathways between literary worlds and students' worlds in the context of the role of education to prepare students 'to meet unprecedented and unpredictable problems' Rosenblatt (1995, 3). The capacities of literature to build bridges between literature, students' lives and society is closely tied to societal challenges, and she places the teaching of literature within democratic education based on Dewey's ideas of democracy 'as a way of life' (Dewey 1916/1959). She claims that teaching of literature has the educative potential to contribute to alternative ways of viewing oneself and society in relation to the imaginary spaces offered in literary texts. However, then the teaching of literature needs to be carried out in connection to students' lives. Rosenblatt (1994, 1995, 2005) advances a reader-response theory of literature based on the concept of transaction (Dewey and Bentley 1949/2008, see also chapters 3 and 10). Reading is here viewed as a transactional process in which meanings of texts are constituted in encounters between readers and texts, and it is this transactive space that constitutes meaning. In addition, literary texts are considered to offer more than literary values (Rosenblatt 1995, 2005), viewing literary depictions as 'environments' that (potentially when encountered) can help readers navigate real-life issues. Thus, literary meaning is understood as lived and shared experience (cf. Werner 2010, 22).

Rosenblatt (1994) presents a practical philosophy of teaching and discusses two different ways of approaching literary texts, *aesthetic* and *efferent*. Taking an aesthetic stance, the attention is paid to 'living through' the text, focusing on the reader's experiences, while in taking an efferent stance, the reader pays attention to deriving information from the text. Rosenblatt (1994) contends that efferent stances have dominated the history of literature education, and she argues for the importance of staging aesthetic stances. If not, the potentiality of bridging transitions between literary events and the lives of the readers is gone missing and, consequently, the teaching of literature loses its educative power as an incentive for individual growth and change and for active participation in society.

Writing

In order to stage encounters with literary texts that open up transactional spaces, the students need to *respond* to what is read. It can thus be argued that it is in the encounter between the *text* and the reader's *response* that meaning is constituted. Thus, staging *educative* encounters with literary texts implies staging appropriate *responses* or *reactions*. Drawing on writing research, the pedagogical principles involve three different aspects of writing: *writing to learn, peer-response* and *learning to write*. *Writing to learn* involves writing activities such as personal reflections, articulations of thoughts and ideas and in combination with reading interpretation of facts, and reflections over diverging perspectives and arguments. Such writing to learn activities take a subject-oriented personal stance to writing, implying that the aim is to enhance personal learning processes that are not assessed (Dyste et al. 2011; Manchón 2011). In comparison, an object-oriented public stance, *learning to write*, focuses on developing writing skills in order to communicate a content efficiently and correctly according to the standards of different genres such as academic writing or poetry writing. Thus, the aim is to learn how to produce texts that are coherent and nuanced, and which are often assessed (Dysthe et al. 2011).

Combining these conceptualisations of writing with Rosenblatt's discussion of aesthetic readings, one can argue that aesthetic readings require a subject-oriented response (i.e. writing to learn) and efferent readings require an object-oriented response. For Dysthe et al. (2011), *writing to learn* is primarily viewed as a strategy for developing writing skills and a strategy for creating participatory and dialogic classrooms. Meanwhile, Rosenblatt (1995) stresses the importance of aesthetic readings as an issue of democratic participation based on the idea that citizens in a democracy need to have the skills to develop their own opinions in relation to other peoples' thoughts and ideas. The issue of responding and sharing responses becomes crucial here. In order to encourage responses to literary texts and to other students' writings, the pedagogical principles discussed in this chapter are inspired by an approach to peer-response (Elbow and Belanoff 1989). They suggest different ways of responding, and the pedagogical principles presented in this chapter involve two of their suggested responding strategies: (1) sharing: no response and (2) pointing and centre of gravity. They are here called *aesthetic responding strategies*, because they imply that the purpose of the response is to 'live through' other students' texts and are thus expressed in a personal language. Thus, responding aesthetically means expressing a response involving personal and subjective reactions to a text. Responding strategies that are more suitable in combination with object-oriented writing are in this chapter labelled *efferent responding strategies*, gearing responses towards a text with the aim of improving the final product, making sure it is written according to external criteria.

One purpose of engaging students to express personal opinions and share responses with others without degrading them is to help them advance understanding for other peoples' experiences, thoughts and ideas that diverge from their own. If the goal of ESE is to empower students to act on sustainability issues in real-life situations, such skills are crucial. Transferring these discussions to ESE, the pedagogical principles suggested in this chapter provide teachers with strategies that can be used to stage literary encounters that target sustainability competences such as *normative competence* (encourage students to express and share conflicting ethical and political values and knowledge uncertainties), *self-awareness competence* (encouraging students' to reflect on personal standpoints, reactions and emotions), *collaboration competence* (sharing responses) and *critical thinking competence* (UNESCO 2017).

The four pedagogical principles

As mentioned above, the four pedagogical principles for staging reading and writing activities in environmental and sustainability education suggested in this chapter have been used and tried out in various educational settings. The principles are to be understood as guiding sources of inspiration for putting read-ing and writing into play with students' experiences. 'Staging' is here used as a metaphorical heuristic for describing a teaching practice where the classroom is viewed as a stage, the students as actors and the teacher as the director and the choreographer of the play which contains the literary texts and the sustain-ability issues to be treated coupled with students' experiences. Of course this is a metaphorical way of describing an approach to teaching sustainable development through encounters with literary texts, but looking back at the example described in the beginning of the chapter, this has been a fruitful way of exploring how to engage students in reading and writing activities in ESE.

The first principle, *getting the actors on stage*, describes strategies for the teacher (the director and the choreographer) to create an attractive *stage* where stu-dents (the actors) can compose manuscripts, practice and perform their plays. This implicates a stage where students are encouraged to encounter, express and exchange experiences with other actors. Quite a lot of time is dedicated to pre-pare students for entering the stage which involves pre-reading activities aiming at preparing students for the actual reading of a literary text. The second prin-ciple, *getting the students into the play*, presents strategies that aim at encouraging students to do something *in* the text, i.e. become *actors* and *participants* in the play by making connections between the literary world and their own worlds. The third principle, *re-viewing the play*, describes how efferent readings can be staged, and the fourth principle, *re-acting the play*, describes suggestions for composing new manuscripts directed at a specific audience with the purpose to be acted out on the public stage.

Daniel Defoe's novel *Robinson Crusoe* is used to illustrate how literary encoun-ters can be staged, putting the four pedagogical principles into practice. *Robinson Crusoe* depicts the story of a man who is stranded on an uninhabited island. As such, the reading can be linked to the real-life sustainability issue of small islands that are particularly vulnerable to climate change, which is also the case for the real Robinson Crusoe island located in the South Pacific Ocean. The main character Robinson is a classical hero of an adventure story, but the novel also abounds with sustainability themes. For example, the novel elicits political and ethical themes such as colonialism and slavery. And, Robinson's existence on the island brings out the issue of human domestication, exploitation and control of nature and animals, and his solitude also evokes personal and emotional themes such as fear, self-reliance, hope, adaptation and creativity.

Principle 1: *getting the actors on stage*

The first challenge involves getting the actors on stage. Students might be reluc-tant to read plays such as *Hamlet* or novels such as *Jane Eyre*, thinking they are too old-fashioned and not of their concern, and this might also be the case when introducing the novel *Robinson Crusoe*. However, getting the actors on stage involves strategies for taking a practical starting point when choosing literary texts and when introducing the themes of the reading. Taking a practical starting point here implies bringing in real-life sustainability issues to create as attractive a stage for participation in the reading as possible rather than departing from the literary world depicted in the novel. When introducing the reading of the novel *Robinson Crusoe* from a sustainability perspective, the introduction can for example depart from the real-world issue of the vulnerability of small islands. In order to evoke and re-activate students' knowledge, thoughts and feelings of the sustainability theme of the vulnerability of small islands, the reading of *Robinson Crusoe* can begin with an aesthetic 'warm-up' writing exercise using an image of a small island as a starting point. Using images when setting scenes for reading gears students' reflections towards the theme but still leaves room for individual interpretations. In relation to the reading of *Robinson Crusoe*, an image of the 'real' Robinson Crusoe island can be a suitable first image in order to create a connection between real-life sustainability issues and the novel.

One important issue for the teacher is to create an attractive stage and positive and comfortable actors who dare and want to enter it. In aesthetic warm-ups, the students need to be informed that the purpose of the writing is to elicit personal thoughts and ideas to introduce a theme, implying that the style of writing (i.e. grammar, spelling and form) is not in the foreground. The students' texts will not be marked or corrected. Rather, the students need to be encouraged not to think too long before they start writing but to 'let their fingers go' and focus on writing flow. For students who get stuck, one strategy to get them going is to encourage them to repeat the last word until they feel they have something new to add. The reason for this quite unorthodox writing style is that continuing to write often encourages students to stretch their thinking through activity. The students are asked to share their texts with each other by reading them out loud, either in pairs or in small groups using an aesthetic response. The purpose of sharing the texts with others is to 'live through' other aesthetic writings as well as to discover other students' perspectives and ideas of the same image rather than commenting on what is missing or what is poorly described. In practice, this means utilising an aesthetic responding strategy by, for example, asking the students to listen to each other and thank the writer for sharing, or to point out something interesting or beautiful in the text rather than utilise an efferent responding strategy focusing on the content of the text. From my experience, many students feel insecure at this point, especially if they are not used to sharing texts with each other. One important aspect for the teacher to get the actors on stage is to encourage the students not to excuse themselves for having written bad texts, reminding them that they are all in the same situation having written a text in a short amount of time, implying it cannot possibly be coherent, poetic or perfectly organised. Again, this is not the purpose of aesthetic writing exercises. Furthermore, one important task here is to remind the students to be encouraging, not to interrupt each other while listening and to spare any irrelevant or pejorative comments, even though opinions and reactions diverge. Often students prefer telling *about* their texts rather than reading them out. But since the purpose is to 'live through' other students' narratives, reading out texts is from my experience more efficient. However, leaving out passages is always fine if students feel uncomfortable sharing parts of a text. Forcing anyone too hard to share is not recommended, since it usually makes it more difficult to get the students on stage if they feel uncomfortable, and we do not want scared or insecure actors. Thus, saying no to share is fine. After the first sharing, the students are asked to assemble the different reactions that are produced in the groups and write a common list of themes which is presented to the class and handed in to the teacher. Thus, the teacher gains information of the students' ideas and thoughts about the theme, their reading, listening and writing skills and how they experienced the sharing and the responding, which support the teacher to develop strategies to get the students into the play. Hopefully, the theme of the vulnerability of small island states is introduced, and if not that will be the task of the teacher using the students' responses as a starting point for introducing the theme.

Principle 2: *getting the actors into the play*

The second principle deals with the challenge of getting the actors into the play. By engaging the students in more or less structured aesthetic readings with the purpose of 'living through' the text, the students are encouraged to write aesthetic responses in their reading journals that deal with what they find interesting and relevant in the novel in relation to their own lives while reading *Robinson Crusoe*. Following the same structure as in the aesthetic warm-up activity discussed above, the students continuously share and discuss their aesthetic responses to the reading in class in order to encourage different opinions, experiences and reactions to take part in the play. If the class is used to engaging in aesthetic readings, a less structured aesthetic reading can involve identifying passages that for some reason catch the students' interest and writing personal reflections during the reading or just writing a personal reading journal, while a more structured aesthetic response can involve handing out questions prior to the reading. One way of sharing journal entries is to use the aesthetic responding strategy of identifying powerful or striking passages that for some reason hit the students when they share their aesthetic responses. As they simultaneously receive reactions on their own choices of passages and their own reflections and reactions, this stage also involves a degree of comparing. To increase awareness of other students' comments, students can mark passages that other students notice as particularly interesting, thoughtful or beautiful in their reading journals, which will increase awareness of how other students respond. This is something that usually encourages and motivates students to participate. The passages from the literary text and the students' aesthetic reactions to the literary text are shared by reading them out loud in small groups. One way of encouraging students to participate in the play is to ask them to choose one favourite passage from each student in each group to be shared with the whole class, making the reactions more public. When the novel is finished, each student summarises the aesthetic reading in their reading journal and hands it in to the teacher.

Principle 3: re-viewing the play

The third principle involves strategies for staging an efferent stance to the reading of *Robinson Crusoe*, which means re-viewing the play with the aim of extracting information from the text that relates to the sustainability themes evoked in the novel. To prepare the students for the efferent reading activity, the teacher builds the bridge between the previous aesthetic reading and the upcoming efferent reading by transforming the students' aesthetic responses into sustainability themes. This can, for example, involve using the students' responses from the aesthetic reading to compose a list of sustainability themes. The efferent reading is introduced by an aesthetic writing activity to elicit the students' own experiences and thoughts of the sustainability themes. The list of sustainability themes is handed out to the students, who are asked to write for a couple of minutes on the themes, eliciting previous knowledge and reactions using the same strategies of sharing and responding as described above. The efferent reading aims at identifying passages in the novel that portray the particular sustainability themes. Here the students can preferably choose which theme they would like to focus on, letting the theme gear the efferent reading. The students are asked to keep an efferent reading journal identifying passages in the text that are relevant for the sustainability theme of their choice accompanied by personal reflections.

Principle 4: re-acting the play

The final principle of *re-acting the play* involves suggestions for composing new manuscripts directed at a specific audience, with the purpose of being acted out on a public stage. In this step, the students are asked to use their previous aesthetic and efferent writings collected in their reading journals with the aim of composing a communicative text which is to be shared more widely with a relevant target group. This will involve further reading of other texts depending on the choice of publication, which includes searching for relevant facts. The students could, for example, write argumentative texts about fear or adaptation that can be sent to the local newspaper, or they could write a report on the present-day state of small islands and include suggestions for how to engage in the issue, or they could engage in fictional writing, transferring the theme of *Robinson Crusoe* to a present-day context. This stage involves moving from aesthetic to efferent reading and writing, and the role of the teacher is to give input on how to communicate the pieces to the target group, giving comments focusing on language and style which implies giving students efferent responses.

Concluding remarks

In this chapter, special attention is paid to the staging of aesthetic readings and responses to literary texts in ESE. It is argued that an aesthetic approach to the teaching of literature can open up spaces where students are encouraged to express, share and respond to sustainability issues as well as to express personal feelings and emotions in relation to such issues through reading of literature in a structured way. As illustrated in the teaching experience in the beginning of the chapter, teachers need strategies for staging aesthetic encounters with literary texts as well as strategies for taking care of the responses that follow, which at first glance can be regarded as innocent and simple, since it just deals with expressing opinions and emotions and reacting to the encountered literature. However, if students are trained to take more efferent stances to the reading of literature, they need to *practice* how to take aesthetic stances. Accordingly, students need to be accustomed to a view of reading and writing as not only being tools for extracting and expressing knowledge about the state of the world but also as means for discovering multiple perspectives of experiencing the world, with the final aim of developing solid and firm opinions that can be used in real-life sustainability discussions. And from my experience, this takes time.

If the ultimate aim of teaching sustainable development is to empower students for future engagement in real-life sustainability issues, aesthetic readings and aesthetic responses are not enough. Thus, teachers will need strategies for making use of the experiences made from aesthetic readings and take them further into new activities for staging efferent readings and responses that promote skills that are needed for such engagement. However, the principles described in this chapter emphasise aesthetic readings due to the complexity of sustainability issues, but also due to the emphasis on empowering active student participation. The principles for encouraging students to express and share opinions and emotions suggested in this chapter acknowledge that doing this requires as much structure and practice as learning how to communicate knowledge in a coherent and structured way. The emphasis of getting the actors on stage by creating an attractive and safe setting in which they are given opportunities to express, test and share ideas, thoughts and feelings without being afraid of making mistakes or being silenced is viewed a prerequisite for getting the students into the play. Hopefully, such experiences will empower students to make their voices heard in future public sustainability debates.

References

Buell, L. (1995) *The Environmental Imagination: Thoreau, Nature Writing and the Formation of American Culture*. Belknap Press of Harvard University Press, Cambridge, MA.

Defoe, D. (2014) *Robinson Crusoe*. Barthelsons förlag AB, Sollentuna.

Dewey, J. (1916/1959) *Democracy and Education*. The Free Press, New York.

Dewey, J. and Bentley, A. (1949/2008) "Knowing and the known". In Boydston, J.A. ed., *The Later Works, 1925–1953, Volume 16:1949–1952*. Southern Illinois University Press, Carbondale, 1–294.

Dunlop, R. (2008) "Open texts and ecological imagination". *Canadian Journal of Environmental Education*, 13(2): 5–10.

Dysthe, O., Hertzberg, F. and Hoel, T.L. (2011) *Skriva för att lära. Skrivande i högre utbildning*. Studentlitteratur, Lund.

Elbow, P. and Belanoff, A. (1989) *A Community of Writers*. Random House, New York.

Garrard, G. (2010) "Problems and prospects in ecocritical pedagogy". *Environmental Education Research*, 16(2): 233–245.

Garrard, G. (2012) *Teaching Ecocriticism and Green Cultural Studies*. Palgrave Macmillan, Basingstoke.

Glotfelty, C. (1996) "Introduction: literary studies in an age of environmental crisis". In Glotfelty, C. and Fromm, H. eds., *The Ecocriticism Reader: Landmarks in Literary Ecology*. University of Georgia Press, Athens, Georgia.

Hansson, P. (2009) "Readings for climate change: Ecocriticism and climate change education research". *Southern African Journal of Environmental Education*, 26: 74–80.

Hansson, P. (2010) "Klimatet och litteraturen – en klimatkritisk läsning av Robinson Crusoe". In Kronlid, D. ed., *Klimatdidaktik: att undervisa för framtiden*. Liber, Stockholm.

Hansson, P. (2014) *Text, Place and Mobility: Investigations of Outdoor Education, Ecocriticism and Environmental Meaning Making*. Acta Universitatis Upsaliensis, Uppsala.

Heise, U. (2006) "The hitchhiker's guide to ecocriticism". *PMLA*, 121(2): 506–516.

IPCC. (2014). "Climate change 2014: Synthesis report". In Core Writing Team, Pachauri, R.K. and Meyer, L.A. eds., *Contribution of Working Groups I, II and III to the Fifth Assessment Report of the Intergovernmental Panel on Climate Change* . IPCC, Geneva, Switzerland, 151 pp.

IPCC. (2018) "Summary for policymakers". In: Masson-Delmotte, V. Zhai, P., Pörtner, H. O., Roberts, D., Skea, J., Shukla, P. R., Pirani, A., Moufouma-Okia, W., Péan, C., Pidcock, R., Connors, S., Matthews, J. B. R., Chen, Y., Zhou, X.,. Gomis, M. I., Lonnoy, E., Maycock, T., Tignor, M., Waterfield, T. eds., *Global Warming of 1.5°C. An IPCC Special Report on the Impacts of Global Warming of 1.5°C Above Pre-Industrial Levels and Related Global Greenhouse Gas Emission Pathways, In The Context of Strengthening the Global Response to the Threat of Climate Change, Sustainable Development, and Efforts to Eradicate Poverty*. World Meteorological Organization, Geneva, Switzerland, 32 pp.

Manchón, R. (2011) *Learning-to-Write and Writing-to-Learn in an Additional Language*. John Benjamins, Amsterdam.

Rosenblatt, L.M. (1985) "Viewpoints: Transaction versus interaction: A terminological rescue operation". *Research in the Teaching of English*, 19(1): 96–107.

Rosenblatt, L.M. (1994) *The Reader, the Text, the Poem: The Transactional Theory of the Literary Work*. Southern Illinois University Press, Carbondale.

Rosenblatt, L.M. (1995) *Literature as Exploration*. Modern Language Association of America, New York.

Rosenblatt, L.M. (2005) *Making Meaning with Texts: Selected Essays*. Heinemann, Portsmouth.

UNESCO. (2017) *Education for Sustainable Development Goals: Learning Objectives*. (https://unesdoc.unesco.org/ark:/48223/pf0000247444). Accessed 8 April 2019.

Wason-Ellam, L. (2010) "Children's literature as a springboard to place-based embodied learning". *Environmental Education Research*, 16(3–4): 279–294.

Werner, B.A. (2010) *Pragmatic Ecocriticism and Equipments for Living*. A dissertation submitted to the Faculty of the graduate school of the University of Minnesota, US.

17 Taking up ethical global issues in the classroom

Louise Sund and Karen Pashby

Introduction

As we head into the second decade of the 21st century, it seems that global issues have never been more pressing. Deeply felt and impactful issues such as forced and unequal access to migration, increasingly severe impacts of climate change and the ever-growing discrepancies between the rich and poor are identified by UNESCO (2002, 2010) as global equity and justice issues.

Education for sustainable development and for global citizenship is included in Goal 4.7 of the UN's Sustainable Development Goals (UNSDGs), which apply to all signatory nations. This involves an obligation to encourage education that explicitly engages with issues of global concern and promotes 'decentering' in an effort to bridge the gap between the local and the global. UNESCO (2014) describes this approach as 'a gradual process of expanding the focus of learners from their local realities to include, connect them to, and provide them with a vision of other realities and possibilities' (UNESCO 2014, 20). A global dimension is also included in the national curricula of many countries. To exemplify, the Swedish national curriculum states that education should provide students with insights that enable them to develop a personal approach to overarching, global environmental issues. Education should also contribute to developing students' sense of international solidarity and responsibility (The Swedish National Agency for Education 2011).

However, applying a global ethical dimension in education is both complex and contentious. This is particularly true when it comes to looking at who defines global problems. 'Global' problems, such as poverty or deforestation, are often perceived by people in the 'Global North' as belonging to countries 'over there'. In the European context, this can create an 'us', who study and solve the problems, and a 'them', who cause the problems and require help. Therefore, deeply ethical approaches to global issues – and pedagogical processes and practices that would contribute to them – are possible only if we recognise the relations of power that have shaped history and engage with critical modes of inquiry.

In this chapter, we offer a didactical reflective tool to support teachers in making choices about content and pedagogy grounded in ethical questions and complex understanding of global justice and equity issues. We then exemplify with some examples from practice and how some teachers take up global issues in their classroom.

Common challenges and opportunities for teaching ethical global issues

While the Sustainable Development Goals serve as a mobilising mechanism to promote engaged scholarship and practice in teaching global issues explicitly in classrooms around the world, it is also important to recognise some key issues. For example, despite the good intentions of UNESCO's work in support of the United Decade of ESD, scholars have raised critical concerns. These include universalising approaches in educational sustainability policies and implicit support for individualism and competition (e.g. Huckle and Wals 2015). Relatedly, scholars in the field of critical global citizenship education have highlighted the ways superficial approaches to global education do not address root causes of global issues and therefore can step over complex ethical issues and contribute to a reproduction of colonial systems of power (e.g. Pashby 2012).

Central to these critiques is the complex nature of pressing global equity and justice issues and the important role of the teacher who presents, frames and engages with these topics in the classroom (cf. Sund and Pashby 2018). Such issues cover a wide range of topics, each with overlapping or competing sets of problems and complex ethical challenges. For example, questions could include: Who has the power to define and to solve the issue? What are the different viewpoints and perspectives on the issue and what are the main points of difference and/or conflict? What have been the main causes and is a historical analysis of the issue offered? As Andreotti (2011) argues, a critical approach puts questions of power up for study, and she recognises that we are all both part of the problem and part of the solution, albeit differently depending on our contexts and positions. For teachers it is important to reflect on responses and take actions to address global equity and justice issues in order to develop a critical stance in their students. Teaching these issues involves responding to a number of challenging questions about the purpose of such education and the values and corresponding pedagogies on which such an approach could be grounded (Sund and Öhman 2014).

Postcolonial theory contributes an understanding that many of the conditions created by colonialism, such as economic inequality, unequal power relations, ethnocentrism and marginalisation, have been intensified by conditions of the current global market that perpetuate the domination of the 'Global South' by the 'Global North'. Seen from this perspective, the history and relationships of colonialism and colonial power influence a global economy based on growth and development that has created and continues to increase poverty while also intensifying environmental issues (Mignolo 2011). Researchers have found that educational initiatives that promote global issues and perspectives, although productive in certain contexts where they may represent a first step, tend to foreclose the complex historical and political nature of global issues and the pos-sibility of more critical approaches (Andreotti 2011; Stein 2018). In line with Andreotti and other researchers, we regard the use of postcolonial perspectives as a critical mode of inquiry that can offer alternative perspectives on international development by challenging ethnocentrism and addressing issues of complicity. However, this scholarship is largely theoretical and has not been translated into resources to be used in educational practice. Andreotti (2011) promotes making use of 'critical literacy' as an educational perspective that identifies and critically examines the origins and implications of assumptions, particularly those underly-ing how problems and solutions are framed. She argues this could prevent the more superficial or 'soft' approaches to global citizenship from reinforcing rather than challenging the colonial systems of power. Andreotti (2012) offers the peda-gogical tool HEADS UP to assist with the task of identifying and addressing historical and present-day colonial patterns of thinking and relationships that can be reproduced by educational initiatives. The acronym comprises the terms: hegemony, ethnocentrism, ahistoricism, depoliticisation, salvationism, uncom-plicated solutions and paternalism. Drawing on postcolonial theory and critical literacy pedagogy, in this chapter we ask in practical terms and as educators, what can we do to push the boundaries of conventional didactic approaches to address root causes of unprecedented global challenges that we currently face?

A didactical reflective tool (DiRe tool) for teaching global issues

Below we have developed and adapted Andreotti's tool into a didactical reflective tool (DiRe tool) with *four key aspects* of a critical engagement with global issues: *contextual-historical, affective, political* and *epistemological*. The contextual-historical aspect addresses whether and how teachers problematise global issues in terms of global injustice and relate them to a historical consciousness in their teaching. The affective aspect takes into account problematisations of benevolence, charity, empathy and responsibility. The political aspect considers how power relations are described in teaching practices and how students can be invited to see themselves as potential agents of change. Finally, the epistemological aspect addresses how teachers deal with a pluralistic perspective of knowledge and to what extent they problematise universal reason in their teaching (Sund 2016).

The DiRe tool supports teachers to reflect on how their teaching approach enables critical engagement with such circumstances as past legacies of colonial-ism and ties to current inequalities, unequal dimensions of power and universal-ising solutions to complex issues. The didactical questions emerging from these aspects can assist teachers to reflect on curriculum planning and classroom prac-tice by raising consciousness as to the extent to which power relations and ethical responsibilities anchored in historical perspective are made transparent in the teaching of global issues. Below are some of the guiding questions tied to each aspect (Table 17.1).

The guiding questions can support teachers to engage with these issues in a generative and ethical way by reflecting on the types of activities, materials and

Table 17.1 A didactical reflective tool (DiRe tool) for teaching global issues

Aspects of global issues	Guiding questions: In my teaching . . .
The contextual-historical aspect	How do I discuss different historical roles and positions in relation to present problems? How can I avoid treating an issue as it happened out of context?
The affective aspect	How can I take up take up altruistic attempts to benefit others without reinforcing an us/them relationship or close opportunities for reducing inequality? How can I navigate between and engage constructively with emotions and sentiments that emerge in processes of teaching justice, responsibility and change?
The political aspect	How can I address inherited and taken-for-granted power relations? How can I invite students to see themselves as potential agents in societal development?
The epistemological aspect	How can I address that there are other logical ways of seeing, understanding and experiencing the world? How can I address people's tendency to want a quick fix?

projects they choose when taking up global issues in their classroom. They can also help teachers to identify what they bring to an issue and underline the need to orientate students to a range of conflicting perspectives. This is important because global issues take on different complexities when viewed from different perspectives based on a temporality (that includes future generations and their needs), history, values, power and knowledge.

Illustrations from educational practice

In the following, we draw on examples from some research we conducted in secondary school classrooms to demonstrate how the DiRe tool highlights key aspects of critically informed educational choices teachers make when teaching global issues, including how they pay attention to the complex and shifting challenges of these issues.

Example 1: global politics and the situation in Syria/the current migrant crises

In the *first* example, Travis, a social sciences teacher, emphasised the *contextual-historical* and *political* aspects. In a course on global politics and the situation in Syria, Travis talked about the wide range of groups involved in the conflict.

Travis: These groups are also, and this is what I mentioned last week so I really want you to pick up on this, it is often presented in the Western media that you have Asad who is generally considered to be a pretty nasty kind of guy fighting against these different groups and that there is a

religious conflict or . . . either that or totalitarian government, a dicta-
torship fighting against pro-democracy *or* against Islamic militancy, or
both . . . And all these things are correct . . . *but* Asad has ruled Syria and
his father before him, a brutal family, about 30 years . . . perhaps even
longer . . . but they are coalition of interests . . . and this reflect *ethnic
tensions* in Syria and we don't hear that much about that in the current
debate . . .

Sana: So everything that goes on in Syria is because of ethnic divisions . . .?
Travis: *No,* you cannot isolate it and say that it is *only* because of ethnic divisions, there are lot of other things that far more complicates it. . . .
Syria is a very complex society, we have this issue going back to Sykes–
Picot Agreement a hundred years ago or so . . . where the boundaries
of Syria were drawn up pretty arbitrary . . . . You have had groups such
as the business class in the city of Aleppo . . . Aleppo is *or was* larger
than Damascus, it was a very large and thriving business community and
these people had their own interests which to *start* with in the conflict
tended to ally with Asad. So there are, if you like, class differences there
as well, religious and ethnic differences, cultural differences underlining
the conflict . . .

When responding to a student question, he expanded on the complexity and
explained the *historical* role of the situation (Sykes–Picot Agreement, the long
reign of the Asad family) and the multi-dimensional and *contextual* circumstances
(religious, ethnic, cultural, class differences) that forms the conflict. Challenging
mainstream perspectives and who is framing the issue (the Western media) as well
as pointing to perhaps less-discussed agents of change (the business class), Travis
also addressed power relations and the *political* aspect of the conflict in his teaching.

Example 2: global development and energy consumption

In the *second* example, a social science course on international relations, the
teacher had instructed students to reflect on the benefits and tensions of global
development and what would happen if India's energy consumption would follow
the same pattern as in the Global North. In their discussion, the two students,
John and Linette, struggled with issues regarding how colonial history continues
to characterise global power relations, but also reflected on how these things
might change over time (see also Sund and Pashby 2018).

John: Statistically you could end poverty tomorrow as long as everyone was
in on it and willing to dole out/distribute all assets . . . but that assumes
that people want to do it also . . .
Linette: But I do not get it when she [the teacher] talks about washing machines
. . . is it that *we do not want* them to have washing machines . . .?
John: Yeah, well look, people think like in India that there are so many people
. . . and if all those people want a washing machine that would heavily

increase energy consumption and it will be an environmental disaster
. . . well that is . . . you cannot argue that way . . . because then you place
yourself above others . . . it is the view of human beings . . . if you think
of that as okay . . . and in that case you hold on to a . . . well you might as
well continue with slave trade . . . this is basically the same principle . . .

Linette: . . . so in that case we should choose not to . . . and that we would not
handle/cope with . . .

John: . . . well in that case we would need to bike more . . .

Relating to one's own view of others ('them', 'those people'), the students
associated the spatial distances between people with relational and *affective*
aspects, i.e. how much sympathy we feel for another group and that we need
to put aside our self-interest ('place yourself above others') and not reinforce
stereotypical roles ('continue with slave trade'). The students also related the
world's growing energy consumption to what we are prepared to change in
our way of living for reducing our consumption of energy. John suggested one
possible behaviour change at the local level, 'biking more', which suggests that
he recognises his potential role as an agent of change for sustainability, thus
emphasising the *political* aspect.

Example 3: sustainable fishing and the nature of knowledge

In the *third* example, the students in science teacher Carla's classroom were
instructed to reflect on sustainable fishing and the fact that not only fish but also
marine mammals are products for human consumption. Carla stressed that it is
a tradition in some indigenous cultures to hunt marine mammals since it is an
important food source, but indigenous people only harvest what they need.

Carla: So, what I want you to discuss is to what extent does our culture determine or shape ethical judgements about ethical issues like this? Because I
could see that some of you . . . you have a *feeling* about this issue . . . And
where does that come from . . .? Would it be different if you were raised
in an Inuit family . . .? What is forming your knowledge about this? And,
how can the rights of indigenous cultures be respected and international
regulations and the rights of species to be protected be reconciled?

In the above example, Carla introduced both an *affective* aspect, that issues are
framed by different values systems, and an *epistemological* aspect by relaying that
how we grapple with these complexities depends on our different experiences of
the world.

Three female students discussed critical reflections on the questions offered by
the teacher:

Sana: I think it is a distant problem . . . something that doesn't affects us *right
now* and . . .'

Melissa: It's *distant*, because it doesn't play a big role in our lives . . .
Sana: That's true . . . not to be rude, but we don't *care* . . . a lot about the whales . . .
Melissa: That's true because they don't *play* . . . they don't *affect us* . . . but for other people . . . not only the *people*, but the whole environment if whales disappear and become extinct or endangered it will affect so much, the other organisms and the oceans . . .
Sana: *Then* we will be the ones to *speak*!
Melissa: When she [the teacher] mentioned indigenous cultures, I don't think that indigenous cultures are the reason why whales are endangered, I think it is like . . .
Sana: No, isn't it more like she that she is talking about that they are relying more on it than we are . . . Or . . .?
Melissa: I don't know . . .
Bisma: I think, we have already mentioned that, like reaching a common ground where we can still fish . . . or *they* . . . indigenous cultures can still fish to a certain extent to make it sustainable . . .
Sana: . . . and let them reproduce . . .
Bisma: . . . respectful . . .
Melissa: . . . but that is so hard to do though . . .
Sana: So sad . . . I think this is the reason why most people become vegetarians . . .
Melissa: . . . but it's not a reason to stop eating fish . . . I don't know . . .
Sana: . . . but if enough people stop eating it, the demand will be less . . . I guess that is how they are thinking . . . the vegetarians . . .
Melissa: Yeah, that makes sense . . .
Bisma: I think it is also about conscience . . . they feel bad about murder or . . .
Sana: I would never be able to become a vegetarian . . . especially because of *my* culture . . .

The three students seemed to agree that ethical and sustainable whaling is something that does not affect them personally because they saw it as existing far away in terms of a geographical, and/or perhaps cultural, distance. They did recognise, however, that the threat of a possible extinction of whales would have a larger scale impact ('the environment', 'other organisms and the ocean') which would make them react.

As the discussion continued, the students touched on the rights of indigenous peoples to engage in their traditional and sustainable fishing practices, showing an understanding that there are other logical ways of experiencing the world and also that there are no easy solutions to harvest and maintain whaling populations and protect endangered species. The students connect the difficult work of respecting indigenous practices, 'so hard to do', with a complex treatment of the rationales for becoming vegetarian, which one of the students also locates as a cultural issue.

In this small discussion, the students highlight the moral dilemma of the possible extinction of a species, in this case whales, its impacts on local and global society and how that would affect them. They also demonstrate a nuanced treatment of the importance of maintaining indigenous peoples' rights to their traditions and the complex intersections with Global North vegetarianism. In weighing various *political* aspects, they demonstrate a respect for the ecological knowledge of indigenous communities, emphasising the *epistemological* aspect. And, they deeply consider an *affective* consideration of different views.

Discussion: what can teachers and educators take from this?

Applying a global ethical dimension in education is both complex and contentious; yet, it is arguably at the heart of education that takes up and responds to the issues contributing to the earth's sustainability crises. The Sustainable Development Goals in combination with curriculum objectives in countries of the Global North open a policy and a practical space for classrooms to engage in learning about global issues, and it is important to use the momentum around the Sustainable Development Goals to fundamentally rethink environmental and sustainability education and avoid repeating the limitations of the past.

In the above examples, teachers and students engage with challenging issues through teaching practices and student interactions. Wals (2015) claims that considering values and ethics in the classroom has been long neglected and that teachers and educators appear uncomfortable with dealing explicitly with values and ethics. This includes reflecting on how the lives we live in the Global North affect the lives of others far away. The students in this study take up opportunities enabled by their teachers to discuss ethical global issues. They are quite aware that global issues are many-sided and concern both environment and social justice, and the teachers take this as a starting point, alongside their own experiences and educational philosophy. These classroom snapshots are examples of how the tool can support reflective approaches to teaching that might start a careful examination of collective 'root' narratives and where such understandings are coming from (Andreotti 2014). Hopefully, such an approach might also enable students to develop awareness about their own values and reflect on the structures of which they are part.

If education aims to develop students' critical understanding of global issues, we as educators need to bring ethical implications of the consequences of global issues into teaching and learning. This is part of engaging in global relations more deeply and reflexively and involves critically engaging with notions of complexity and complicity as part of the daily life of classrooms.

As we begin to implement practices in support of Sustainable Development Goal 4.7, it is essential that we engage directly with the ethical issues at the root of sustainability issues. This will require pedagogy that involves both hope for a better future and deep concern with the realities of today's inequalities. These examples demonstrate that teachers and students can experience a sense of significance and worthiness of engaging in a more critical approach. If we fail to engage with ethical global issues pedagogy, we run the risk, as pointed out by Andreotti (2006), of unintentionally contributing to the issues we are trying to solve. And if we critically reflect and support students to do so, we open up possibilities for approaches to global issues pedagogy that come much closer to addressing the pressing issues of our deeply unequal world.

References

Andreotti, V. (2006) "Theory without practice is idle, practice without theory is blind. The potential contributions of post-colonial theory to development education". *Development Education Journal*, 12(3): 7–10.

Andreotti, V. (2011) *Actionable Postcolonial Theory in Education*. Palgrave Macmillan, New York.

Andreotti, V. (2012) "Editor's preface: HEADS UP". *Critical Literacy: Theories and Practices*, 6(1): 1–3.

Andreotti, V. (2014) "Critical literacy: Theories and practices in development education". *Policy & Practice. A Development Education Review*, 19: 12–32.

Huckle, J. and Wals, A.E.J. (2015) "The UN decade of education for sustainable development: Business as usual in the end". *Environmental Education Research*, 21(3): 491–505.

Mignolo, W. (2011) "The global south and world dis/order". *Journal of Anthropological Research*, 67(2): 165–188.

Pashby, K. (2012) "Questions for global citizenship education in the context of the 'new imperialism'". In de Oliveira Andreotti, V. and de Souza, L.M. eds., *Postcolonial Perspectives on Global Citizenship Education*. Routledge, New York, 9–26.

Stein, S. (2018) "Editorial. Rethinking critical approaches to global and development education". *Policy & Practice. A Development Education Review*, 27: 1–13.

Sund, L. (2016) "Facing global sustainability issues: Teachers' experiences of their own practices in environmental and sustainability education". *Environmental Education Research*, 22(6): 788–805.

Sund, L. and Öhman, J. (2014) "On the need to repoliticise environmental and sustainability education: Rethinking the postpolitical consensus". *Environmental Education Research*, 20(5): 639–659.

Sund, L. and Pashby, K. (2018) "'Is it that *we do not want* them to have washing machines?': Ethical global issues pedagogy in Swedish classrooms". *Sustainability*, 10(10), 3552, 1–13.

The Swedish National Agency for Education. (2011) *Curriculum for the Upper Secondary School*. Fritzes, Stockholm.

UNESCO. (2002) *Education for Sustainability, From Rio to Johannesburg: Lessons Learnt From a Decade of Commitment*. UNESCO, Paris. (http://unesdoc.unesco.org/images/0012/001271/127100e.pdf). Accessed 21 January 2011.

UNESCO. (2010) *Education for Sustainable Development in Action Learning & Training Tools (2)*. Section for Education for Sustainable Development, Paris. (http://unesdoc.unesco.org/images/0019/001908/190898e.pdf). Accessed 5 April 2018.

UNESCO. (2014) *Global Citizenship Education Preparing Learners for the Challenges of the Twenty-First Century*. UNESCO, Paris. (http://unesdoc.unesco.org/images/0022/002277/227729e.pdf) Accessed 5 April 2018.

Wals, A.E.J. (2015) *Beyond Unreasonable Doubt Education and Learning for Socio-Ecological Sustainability in the Anthropocene*. Inaugural address held upon accepting the personal Chair of Transformative Learning for Socio-Ecological Sustainability at Wageningen University on 17 December 2015. Wageningen University, Wageningen (https://arjenwals.files.wordpress.com/2016/02/8412100972_rvb_inauguratie-wals_oratieboekje_v02.pdf). Accessed 5 May 2018.

18 Students as political subjects in discourses on sustainable development – a glimpse from Sarah's classroom

Iann Lundegård

Introduction

It's late afternoon. Darkness is falling and large snowflakes float slowly down outside the classroom window. Sarah will soon teach her high school class, consisting of 16 students. One after another they fall into the classroom, now tired after several heavy lessons earlier this day. Soon Sarah will have had this group for a whole semester. She knows them quite well. They are in grade 2 in upper secondary school and in different programmes in their education, but as an individual option, they have all chosen the topic of 'global environment', an area they want to immerse themselves in. Sarah herself has an academic background in science and philosophy. It's a combination of two subjects that she believes fertilise each other in a positive way. It is also in these two areas where she mainly teaches in school. Now she is responsible for this optional course, which she designed herself.

In this chapter, we will have a glimpse from a classroom in which the teacher has the aim of organising an authentic education in the sense that it allows the students to act as political subjects. From a didactical perspective, it describes how a teacher directs a teaching that acknowledges the inevitable link between knowledge and values. Furthermore, with the help of a pragmatic perspective, we can see how these two aspects of the students' meaning-making closely interact. Simultaneously, we get a deeper insight into how an ESE education based on freedom and democracy render opportunities for the students to declare personal relations to the content at hand, but also how this freedom is ultimately achieved through mutual encounters in-between them (see chapter 5). The chapter has gained inspiration from a previously published scientific study in this area (Lundegård 2018).

Teaching within a context of values

When Sarah starts her teaching this day, it is with an idea that her students are here not only to comprehend the subject that comes up during the lesson. She also wants them to have the opportunity to reflect and take a stand based on values concerning the arising issues. Thus, they will have an opportunity to realise how this knowledge brings relevance to them in their lives. These thoughts are by no means random. Sarah knows that most of the content students encounter and the skills they acquire in school are, whether we want to realise it or not, loaded with values (see chapter 20). That everything we call knowledge at some point is developed for specific purposes, in a process that has consequences for people in their real lives. That the classical division between knowledge and values is just a construction originally made in order to make it easier to carry out an everyday conversation, but which through time, more or less, has become cemented as scientific truth. She has brought this insight from her own background in philosophy, but also from teacher education. By making this connection seriously and then bringing in knowledge and values as a common package in teaching, she not only raises questions relevant to the students in the future, but also creates an education that is authentic for them here and now (Lundegård 2018), i.e. a teaching that encourages them to want to join. For real!

Authenticity is often discussed in scientific teaching, especially when related to environmental and sustainable development issues. It has more or less become a buzzword in these areas. It is often used in discussions about students' association to context and affiliations to place, but just as often dealt with as a state of consciousness, or frame of mind. Murphy et al. (2006) address authenticity as either cultural or personal. Thus, *cultural identity* is reached when school activities correspond with other activities within society. *Personal authenticity* on the other hand is identified when learners have the ability to perceive value and meaningfulness in teaching. For Sarah's part, both entrances are important, i.e. that her teaching raises issues that are in the news in society while simultaneously trying to create a teaching methodology that invites students into a situation where emotional utterances are given space and where students have the opportunity to act as political subjects. Don't put pressure on, but invite. It's important!

The educational and pragmatist philosopher John Dewey (1893), who in several ways has had a great influence on Sarah's ideas on philosophy, pointed to what he thought was an educational fallacy, namely that an on-going event taking place here and now is meant to be carried out mainly in order to achieve some learning of what can be useful elsewhere and later in life. Even these days, it is argued that schooling is estranged from the students' reality and lacks personal meaning. Most often, they are expected to take part in order to achieve generic knowledge abilities and competencies required elsewhere or further on. Thus, particularly when the students are expected to engage in contemporary issues and problems happening globally and in society here and now, this discussion of authenticity is, without doubt, of significant importance. That's also why Sarah wants to create an education that is close to the students. Today she does it by working with values-clarification methodology.

The values-clarification methodology

Values-clarification exercises are something that students all over the world today encounter in different ways. In such exercises, they have to change chairs or stand in classroom corners, for example, in response to statements or options of action offered by the teacher. The methodology has been around for a long time. It was first created in the late 1950s by Louis Raths and subsequently developed by, among others, Sid Simon at the Values Realization Institute at the University of Massachusetts (Kirschenbaum et al. 1977). Both of these educators were disciples of John Dewey, and the methodology links directly to a profound discussion of democracy as pivotal in education. Actually, this method is more interested in the democratic process than in spreading a specific democratic value. The whole discussion begins with a pragmatic assumption that there is no universal or absolute good or right and true (see chapter 20). Ultimately, neither God, the market, nature or anything else determines the right way to act. Decisions about which paths we will choose are settled between people in on-going democratic processes.

Through values-clarification exercises, a teacher can help students to realise that there are always choices, as well as help them create a personal awareness of the choices they make (see chapter 13). This method has often been used in teaching about drugs, bullying, sexuality and relationships, and in other contexts where the content is close to students' personal values. In order for a valuation to be active, it is said, it is simply not enough to declare it for yourself. You have to visualise your choices for others. It is by mirroring yourself directly into the eyes of others that you emerge as a free individual. Thus, the crucial point with the exercises is not to get the students to identify what is right and true in absolute terms, but instead to help them clarify what values and choices they think best suit them. In this way, students will not only acquire knowledge from teaching that will help them deliver the correct answers on an exam, but will also be able to directly influence things developing around them. Sarah knows that this approach is successful in her teaching, especially when it comes to issues related to the future and sustainability.

The actual assignment

For the present, Sarah has partially turned things around. Instead of creating the allegations in the exercise to take a stand for herself, she has she handed it over to the students. Now she has divided the class into four groups that, under her supervision, have chosen a specific area related to sustainability issues in which to immerse themselves: gene technology, climate change, resource distribution, etc. The actual task they have received before the lesson is to briefly present to each other what they came up with in their group work. A more detailed written report will later be submitted to the teacher. But what they have also been asked for is to create a values-clarification exercise for their classmates to participate in. Also, the entire assignment has been preceded by an exercise in which Sarah gave instructions on how to carry out such an exercise.

About countries' food supply – on a factual level

The first group to make their presentation, led by John, has deepened their knowledge in the field of sustainable food production. They begin with a short presentation about the ecological aspects of food supply, energy loss in different nutrition chains and the significance of the earth's climate zones for what is possible to grow on different parts of the planet. During their search for knowledge, they have also been confronted with different governmental policies on production and trade. This research has brought them to their current thoughts and doubts about how the planet's resources should be distributed in a fair way and is an issue the group wants to raise and discuss with their peers. The group has chosen to carry out an exercise in which their classmates have to choose between two options. John takes the floor and asks, 'Agriculture now supports twelve billion people. How can we create a fair distribution of the resources on the planet?'

He continues by giving the participants two options to choose between: 'Should we remove duties, or should we instead give further assistance to local agriculture?'

After a few seconds of thought, the rest of the class begins to move towards two opposing corners of the classroom. Some of them a bit slower and others with determined steps. When the movement has died down, one of the students, Kim, signals that he represents and wants to argue for the option of removing duties between countries.

John turns to him and asks, 'Why would this be a better option then?'

Kim replies with a strong voice, 'Because then you can just transport food to those countries that need it.'

Now several of the listening students start to wave their arms to reply to this, which results in Kim looking down at the floor, muttering to himself, 'But what's bad about it then?'

A third student, Jayden, takes over and says, 'But then the countries will depend on each other'.

And another student, Alfie finishes Jayden's utterance by adding more substance to the issue. 'That's not the problem maybe. The problem is that there are such different economies. Cambodians really can buy cheap Swedish food, because food's so expensive because there's quite an expensive labour here', he says. Based on something he has seen on TV, he continues to argue. 'Moreover, it has been shown that if you transport cheap food to developing countries, it affects their production in a negative way. I saw a program, where a company transported cheap powdered milk to a country and the dairy farmers in that country couldn't compete so, when you think that it's really good to offer cheap food, to developing countries, it can have negative consequences.'

What at a first glance looked like a simple question here became ever more complex. When the students realised that since different countries have different economies, the removal of duties might render more trouble.

Sarah thinks that this became a really interesting discussion in which students were able to twist and turn the problem and present factual arguments more than just opinions. She also notes that her students learned a lot about sustainable development during their work. Above all, this group, who gained more knowledge in the particular field, could make constructive follow-up questions and provide additional facts that deepened the discussion. This also resulted in the dynamic that those who participated in the exercise eventually changed their positions because they were convinced of the opposite. That makes Sarah happy.

A concept connected to this context and that has been debated on a comprehensive level is pluralism (Rudsberg and Öhman 2010; see chapter 5). In pluralistic teaching, students are given the opportunity to identify the conflicts of interest involved. Furthermore, an education in the spirit of pluralism means that the students already in teaching are trained to participate in democratic deliberation with others on public issues (Van Poeck and Vandenabeele 2012). However, to experience pluralism is not just to visualise what you already know. In teaching characterised by pluralism, students are also given the opportunity to pay attention to perspectives that may not always be immediately apparent. Within a critical or radical pluralism (Sullivan 2001), marginalised, often-excluded perspectives on the issues are highlighted.

In this values-clarification exercise, it was obvious that the students took a stand. Using their bodies, they positioned themselves in relation to the highlighted conflicts. But what strikes Sarah at this point is that, apart from the students being placed into the respective corners of the exercise, none of them fully exposed him/herself in relation to each other. There was simply no such kind of statement that they had to face. Obviously, they took a position on the matter at hand and argued mainly based on facts, but they had difficulty with drawing largely from a personal level. Actually, this is not any sort of demand or expectation that Sarah has made, but sometimes she wants her teaching to offer more than just this factual negotiation. She wants her students to have the opportunity to try each other in an immediate verbal exchange of views.

In this discussion about the relationship between facts and values, it is interesting to look at how the relationship between the two has been discussed philosophically. When, for example, the philosopher Ludwig Wittgenstein (1967, 1969) made his investigations into how people perform in communication, he often returned to the term 'language game'. This is also a recurring concept among philosophers who succeeded him and struggled with questions about how people communicate. One of these is the philosopher Stanley Cavell (1979/1999), who was primarily interested in, and identified, two overall prevailing language games in human communication. In the first, which he refers to as 'epistemological', we relate to facts and criteria. In the other, he called 'moral', reason is claimed from evaluating statements. A natural scientist, for example, is often expected to argue epistemologically in a way that is based on causalities, and regularities are sometimes summarised in theories. In an epistemological language game, we make the world, and phenomena within it, understandable from interconnected knowledge valid within certain framework based on criteria. What Cavell (1979/1999) refers to as a moral language game, on the other hand, is often present when we take a stand on something during our communication in ways that do not explicitly make use of criteria. When, for example, people express value judgements such as, 'it is unfair', or 'how beautiful', we usually don't ask for logic or require any further investigation. Instead, we give credit to a person who decides that the utterance is valid for them, although we may not come to the same conclusion within our own experience. What is important in this context is not to see this division between the epistemological and moral as a solid dictum of domains in the world and what our communication ultimately represents. Such has never been Wittgenstein's or Cavell's ambition. Rather, the whole should be seen as an attempt to study how we carry on in communication and the actual consequences of this. Thus, when we stay in an epistemological language game, we treat the content in impersonal terms. However, when we make use of terms associated with values and emotions, this will function as an immediate declarations of our own position (I-relation) to the issue at hand (Lundegård 2008). But this is not to be confused with a report on an inner feeling. Rather, when we in an immediate language game declare ourselves in terms like 'I love', 'I want' or 'I deny', it is a performative speech act (Sullivan 2001), enforced in relation to those whom we encounter face to face. Not to represent an inner state, but to anticipate and cause oneself in action. However, just to dare to enter the stage and create a position to be responsible for in a societal context is obviously costly. That is why this performative speech act constitutes what here is referred to as a declaration of oneself as a 'political subject'.

To summarise: 'The political subject' emerges in the very moment one individual takes a performative position to a conflict of interest noticed in the public sphere, outlined in the immediate language game with 'the other' (Lundegård and Wickman 2012; Lundegård 2018).

Here Sara could see that in the first values-clarification exercise, her students actively responded using their bodies as tools. They first marked a position toward the claim they met and then tried to substantiate it with an epistemological language game. But, in order to 'become someone' – individually visible in relation to the content – Sarah thinks it is required that the students are allowed (not forced!) to meet each other directly as political subjects.

About laws and activism – the student as a political subject

A bit further into the lesson, another group presents the area they have delved into, the increasing greenhouse effect, global warming and climate change. In this work, they were, of course, given a thorough understanding of all science-related cause-effect relations that associate the climate change problem with fossil fuel combustion. However, the main question might rather be how they can deal with that knowledge and what consequences it brings to their own actions in the future. When the students in that group reported the facts they gathered for their classmates and convinced them about the scientific links, they asked, just how far you are willing to go if you want to act based on these consequences? In connection with the gathering of facts, the group also encountered information on how different environmental organisations have chosen to act in order to curb the spread of climate devastation.

The valuation claim that the group puts in front of their friends is pronounced by the leader of the discussion, Emma, 'It is okay to break the law in order to save the environment?'

Once again, the students have chosen to perform a simple exercise with only two corners, 'yes' in the one and 'no' in the other. Most of the students soon slip into respective corners, while some who don't choose to take a stand directly linger instead as spectators, which also is quite acceptable in such a values-clarification exercise. The discussion takes off, and eventually it comes round to different activists' law-breaking actions in which they, among other things, slashed the tires of city jeeps. It bounces back and forth between those who are for and those who oppose the illegal actions. Linn and Harry, who support the actions, argue that breaking the law may be justified for the sake of the environment. Linn gesticulates broadly with her arms as she advocates two arguments, 'I think the laws, now drawn up for the environment is pretty lame', she says and continues, 'So I *can really understand* if you don't want to respect them. There are actually quite few laws that take care for the environment.' Within this statement, she directly addresses one of the other students who previously claimed that the laws have actually come about through democratic deliberation, 'and therefore should be respected'.

'And then also', says Linn, and nods in the direction of one of the other. 'You say that one better should discuss and debate the question. There is no actual discussion about it, and that's the reason why organisations went to such extreme actions. To create a real debate! *Which I think is extremely good! I may not think* the actual action is so very good, when they shred car tires on city jeeps, but it generates a debate. *And that's good, I think.*'

Her classmate Harry agrees. He as well looks straight into the others' direction when he confirms and continues Linn's judgement. 'Sure laws exist for a good reason, but that reason can be over a hundred years old. It can take a very long time to change a law.'

Then he continuous to talk about the slowness of forming the laws and about what he identifies as the conservative system in society. He sets corporate power in relation to the peoples' idealistic struggle. Eventually makes his point. '*That's why I think it is such a bad argument* that you should obey an old law just because it is stated.'

Harry challenges the overall system of regulations that was established long before he himself took place in the public sphere, and continues emphatically. 'There aren't that many laws that I have been involved in and influenced. It depends on what view you have on the government we have. *I think it's so silly* that a law . . . I *can't really understand such an argument.*'

Linn who has been silent for a short while listening to Harry now turns confirmatory towards him, even more eagerly gesticulating with her arms and continues. 'In this case you'll be punished only because you have a specific feeling and commitment. *And that's what I think is absurd.*'

Linn is sincerely pissed off! Her point of view does not match the position some of her classmates defend as postulated by previous generations. She flags this through immediate personal reactions and thus a political stance in encounter with the others. At the same time, she and Harry continuously try to find epistemological arguments that support their position, arguments about the laws being old and out-dated. Sarah, who encountered this kind of argument earlier in similar contexts, realises that this whole discussion also links to an overall philosophical question about the individual's possibilities and rights to free him/herself from the collective. A recurring issue in the context of sustainability issues is how to evaluate the individual's freedom compared to laws and regulations created by societal norms. The whole paradigm of sustainable development has sometimes been criticised just because it attempts to push people from all corners of the world into a common view of what is to be protected and how it should be implemented. This renders questions like: 'Where does the limit of the individual's freedom occur?', 'Whose views should be included in the existing order?' and 'In what way should historical decisions stand aside for new conditions and the next generation's objects of responsibility and care?' These contradictions are elaborated in different contexts and thus referred to as a tension between, for example, promoting subjectification versus advocating universalism conducted by socialisation (Biesta 2010; Lundegård 2018; Todd 2009, see also chapter 4).

Freedom by mutual dependency

A political philosopher Sarah often returns to in this context is Hannah Arendt (1958/1977), who always estimates human freedom in relation to our essential mutual dependency as human beings. Someone who is socially sovereign, she argues, can never perform freely, putting things in motion and bringing something new to the world. Thus, sovereignty and freedom appear as each other's opposites. People achieve true freedom first as situated in networks of relationships/transactions. Or this might be better explained by Sullivan (2001, 19): 'it is able to come into being only because of the way in which the world, including other human and non-human organisms, permeates and in turn is impacted by it'. This means a freedom where my expression appears in the immediate reflection of yours, and always in relation to something we, without further reflection, essentially agree upon. Thus, it is in this mutual encounter with the others that we paradoxically form ourselves as individuals and political subjects (Lundegård 2018). Consequently, that's basically the experience the students have had the opportunity to encounter under the guidance of Sarah's value-loaded teaching, she thinks.

Now it's really dark outside the classroom windows. The snow has diminished. The students slide out. Their steps seem a bit more relaxed now than when they came in an hour earlier. At least some of them continue to gesticulate and discuss when they pass through the door. It is in such a situation that Sarah is sure about the value in continuously developing a teaching where the students' knowledge acquisition is closely linked to issues that are authentic to them.

References

Arendt, H. (1958/1977) *The Human Condition*. University of Chicago Press, Chicago.

Biesta, G. (2010) *What Is Education For? Good Education in an Age of Measurement: Ethics, Politics, Democracy*. Paradigm Publishers, Boulder.

Cavell, S. (1979/1999) *The Claim of Reason. Wittgenstein, Scepticism, Morality and Tragedy*. Oxford University Press, New York and Oxford.

Dewey, J. (1893) "Self-realization as the moral ideal". *The Philosophical Review*, 2: 652–664.

Kirschenbaum, H., Harmin, M., Leland, Howe, L.B. and Simon, S.B. (1977) "In defense of values clarification". *The Phi Delta Kappan*, 58: 743–746.

Lundegård, I. (2008) "Self, values and the world – Young people in dialogue on sustainable development". In Öhman, J. ed., *Values and Democracy in ESD – Contribution From Swedish Research*. Liber, Stockholm, 123–144.

Lundegård, I. (2018) "Personal authenticity and political subjectivity in student deliberation in environmental and sustainability education". *Environmental Education Research*, 24 online.

Lundegård, I. and Wickman, P.-O. (2012) "It takes two to tango: Studying how students constitute political subjects in discourses on sustainable development". *Environmental Education Research*, 18: 153–169.

Murphy, P., Lunn, S. and Jones, H. (2006) "The impact of authentic learning on students' engagement with physics". *Curriculum Journal*, 17: 229–246.

Rudsberg, K. and Öhman, J. (2010) "Pluralism in practice: Experiences from Swedish evaluation, school development and research". *Environmental Education Research*, 16: 95–111.

Sullivan, S. ed. (2001) *Living Across and Through Skins: Transactional Bodies, Pragmatism, and Feminism*. University of Indiana Press, Bloomington.

Todd, S. ed. (2009) *Toward an Imperfect Education, Facing Humanity, Rethinking Cosmopolitanism*. Paradigm Publishers, Boulder.

Van Poeck, K. and Vandenabeele, J. (2012) "Learning from sustainable development: Education in the light of public issues". *Environmental Education Research*, 18: 541–552.

Wittgenstein, L. (1967) *Philosophical Investigations*. Blackwell, Oxford.

Wittgenstein, L. (1969) *On Certainty*. Blackwell, Oxford.

19 Embodied experiences of 'decision-making' in the face of uncertain and complex sustainability issues

Pernilla Andersson

Introduction

While writing this chapter in spring 2018, Dow Chemical is suing the French government for banning one of the corporation's bestselling pesticides. The French government's decision is supported by research regarding risks for bees associated with the use of neonicotinoid pesticides (Rundlof et al. 2015). Both sides defend their positions with reference to food security. The French government (and the European Union) argue that the use of neonicotinoid pesticides is a threat to the well-being of animals, humans and biodiversity, and Dow Chemical claims that the research is based on only a very limited number of bee species, that the risks are exaggerated and banning neonicotinoid pesticides therefore is a threat to economic growth and the well-being of humans. At the same time, scientists know it would be an insurmountable task to include all bee species in experiments. This example illustrates a sustainability issue characterised by complexity and uncertainty and how such issues imply severe challenges when relying on science as a source for reliable knowledge when making decisions (either as a member of parliament, government official, judge, business owner or executive manager in a business). This 'wicked' character of sustainability issues points to the need for the capability to make decisions in the absence of previously established guidelines (see also chapter 2). Accordingly, education needs to prepare students to make decisions in the absence of clear guidelines, formal regulations and legislation, students who in the future could be working in corporations making decisions about what products to sell or not to sell.

Furthermore, relating to how education needs to prepare students to contribute to sustainable development, it has been a debate within the field of environment and sustainability education research that has been described as a tension between instrumental and emancipatory educational objectives (Wals 2010). On the one side, it has been argued that the urgency to combat climate change, for instance, calls for more instrumentalist teaching approaches (Kopnina 2012), while others believe that such approaches are incompatible with emancipation and critical thinking as educational objectives (Biesta 2014). In addition, it has been argued that we need to rethink the post-political consensus and re-politicise environment and sustainability education (Sund and Öhman 2014). This points to the need for teaching approaches that offer space for deliberation of different and conflicting perspectives on what a sustainable future might mean, and possibilities for the students to practice making decisions 'for sustainability' in the absence of clear guidelines without compromising emancipatory education ideals. The aim of this chapter is to address this need by presenting teaching approaches that could offer students these opportunities. The teaching approaches could be combined with each other or used separately. Short practical examples from teaching business economics in upper secondary education are provided to specifically illuminate (a) when different worldviews or perspectives on sustainable development come to the fore and (b) emancipatory educational qualities in terms of subjectification (Biesta 2014; see also chapter 4). The last section presents a didactic model drawing on the concept of 'dislocatory moments' that could help identify room for subjectification and 'business as un-usual' in classroom practice. Subjectification is here understood as a process where subjectivity (which includes affect) is facilitated and involved, and 'business as un-usual' means a change of logic or guiding principle (Glynos and Howarth 2007; Howarth 2013).

'Four corners' – to experience the political dimension of sustainability issues

'Four corners' is a values-clarification exercise (see also chapter 18) where students are invited to take a stand regarding a complex or controversial issue by moving physically in the classroom. The exercise offers students opportunities to listen to different perspectives, verbally communicate their own perspectives and think critically. Accordingly, the exercise can be used as a way to clarify different and conflicting perspectives on sustainability issues. In this way, it offers a way to 'politicise' approaches to address sustainability issues. The exercise is described step by step in Textbox 19.1.

Considering the aim of providing students with an opportunity to 'clarify'[1] their own values in relation to other values or contesting perspectives (similar to conflict-oriented deliberation as described in chapter 8; see also Håkansson et al. 2018, 105), the exercise is more fruitful when students choose different corners. So how can a teacher contribute to this? The alternatives need to highlight or point to the complexities and not be perceived as 'obviously correct or incorrect'. What this means could differ between different groups of students, so the formulation or fine-tuning of the alternatives is something that the teacher must take responsibility for. It is also important for the teacher to contribute to conditions where everyone feels safe to express their thoughts and opinions without being disrespected. This could mean clarifying rules about listening to and not interrupting each other or giving disrespectful comments. This also means being attentive to and responding to expressions of opinions that might cause harm to others. Students standing alone in a corner could be particularly exposed and are therefore worth specific attention. I will come back to this in a practical example later in this section.

Textbox 19.1　Four-corner exercise – step by step

Step 1: presentation of statement or question

The classroom is arranged so that there is empty space to move between the different corners. The teacher presents a controversial statement or asks the students a question. The teacher invites the students to take a stand by choosing one of four corners in the classroom. Each corner represents an opinion or a response, three corners have answers pre-formulated by the teacher and the fourth corner represents an 'open choice'. The teacher clarifies that it is ok to change corners later if they change their mind.

Step 2: students' choosing corner

Students make their choice without any prior discussion and are then asked to physically move to one of the corners. To minimise the risk of students acting in accordance to group pressure, it is important that everyone make up their mind in silence and that nobody starts to move until everyone has made their choice.

Step 3: talking and listening

The teacher encourages the students to elaborate their choice of corner, while also respecting the students and emphasising that it is ok to just take a stand without explaining why.

Step 4: changing corners

When all students have had a chance to develop reasons for their choice of corner, the teacher asks whether anyone would like to change corner as a response to what has come up from the other students' reasoning. Students who want to change corner do so and the teacher offers them the possibility to elaborate what made them change their mind.

For further reference, see for example: www.theteachertoolkit.com/index. php/tool/four-corners

In the context of a market liberal discourse, it is assumed that trade always leads to socially and environmentally optimal outcomes through a 'trickle-down mechanism' (Dryzek 2013, 123). From this perspective, trade is regarded as desirable/good in itself (i.e. sustainable). To politicise approaches to address sustainability issues in a context where students are familiar with or could be invested in such a discourse could therefore involve challenging the notion of trade as good (sustainable) in itself. One way to do this is to raise questions about the relations between global trade, sustainable development and responsibilities. This could be done in the form of a four-corner exercise by posing statements such as: (a) *A corporation should inform customers about sustainability and aim to change their behaviour to become more sustainable* (1. Yes, corporations have an important role in influencing customer behaviour, 2. No, a corporation should not impose values on customers, 3. No, a corporation should only aim to make high profit, 4. Open corner); or asking questions such as: (b) *Should corporations have operations in all countries?* (1. Yes, 2. No, only in democratic countries respecting human rights, 3. No, we should strive to keep all production within national borders, 4. Open corner); or (c) *How could income gaps between rich and poor countries be reduced?* (1. Fair trade, 2. Increased trade, 3. Reduction of trade and increased self-sufficiency, 4. Open corner). The following practical example illuminates how different and conflicting perspectives relating to trade and sustainability can come to the fore when inviting students to take a stand on question (c).

It was a lesson about global trade and sustainable development. The 17-year-old students had watched a documentary about environmental degradation, poor working conditions and suffering of animals related to global trade before the teacher introduced the four-corner exercise. The teacher asked the students to take a stand regarding how the income gap between rich and poor countries could be alleviated by positioning themselves in one of four corners. All but one student in the class moved to the 'fair trade' corner and one student moved to the 'increase of trade' corner. The teacher invited the student standing alone in the 'increase of trade' corner to share her thoughts, and she answered hesitantly:

Student 1: I am a little bit confused but I think that if we increase trade in the world and everybody is making trade with each other maybe . . . [silence]

Teacher 1: . . . that it [trade] is an end in itself?' Yes, it is what it is like to today, we have [trade] organisations with the aim to increase trade and they have no environmental goals in their 'politics'.

Student 1: Hm, yes.

So, how can we understand Student 1's reasoning in the above quote? To do that we need to take the broader context into consideration, which here is teaching and learning business economics and that it is reasonable to assume that the students are acquainted with a market liberal discourse. From such a perspective, the student's hesitation and explicit expression of being confused (or, sign of being dislocated) could be understood in that it does not make sense to talk about trade as potentially unsustainable when trade is regarded as desirable in itself. This interpretation is supported in that the teacher responds to the student's hesitation and silence by filling in 'that it is an end in itself?' This 'confirming' response from the teacher could also be described as a way of caring for the student who here takes a (social) risk when taking a stand no other classmate is taking. The student's answer could be confirmed in that it indeed reflected a common standpoint.

The situation above unfolded a conflict regarding what should be expected from someone (here a business or corporation) acting 'sustainably'. In other words, a situation where it is no longer clear how to proceed or 'go on'. Should one as a person working in a business focus on just 'doing business' or not? This kind of situation is double-sided in that it involves a *loss* of a guiding principle at the same time as the process of finding a new guiding principle offers some *freedom*. In the lessons following this exercise, the teacher supported the students' process involving 'finding new guiding principles' by arranging different situations that allowed the students to explore different perspectives of the role of business in relation to sustainability without explicit instructions of a correct way. In the final section, I will come back to this example to further illuminate in what way this process could be described in terms of subjectification.

Embodied explorations of decision-making with Forum Play

Forum Play is a variation of Forum Theatre developed as a method for the particular use in educational settings by the Swedish drama teacher Katrin Byréus (2001, see also Österlind 2011). Forum Theatre was originally created by Augusto Boal (2008) as part of what he called 'Theatre of the Oppressed' after Paulo Freire's writings on oppression. It was designed as an opportunity for people's collective rehearsal and embodied practice of how to change their world. As developed by Byréus, Forum Play can be described as a kind of role-play with the purpose to process injustices and conflicts between groups and individuals. The audience (which Boal referred to as *spect-actors*) are invited to take active part and reflect about different individuals' alternative courses for action, and the aim is to explore how oppressed individuals[2] can have an influence in particular situations. Here, I suggest a teaching approach inspired by Forum Play that could facilitate a space where students can experiment and explore different strategies or decisions to strive to turn something found unsustainable into something more sustainable. To show how the exercise could be made relevant for a particular group of students, business students are used as an example in the step-by-step instructions below. The choice of example is particularly relevant in relation to the introductory example and for students who in the future could be working in corporations such as Dow Chemical and making decisions about what products to sell or not to sell.

There are different lines of thought when it comes to the role of business in relation to taking responsibility for sustainable development. One line of thought is represented by the expression 'the business of a business is to do business' coined by Milton Friedman. This position represents what could be described as an a-moral or apolitical business ideal. Another line of thought is that a business, apart from profit, needs to have other goals involving, for instance, contributing to the common good (Porter and Kramer 2011), a position that could be described as a 'political' or 'sub-political' business ideal (Beck 2009). 'To be or not to be political?' is thus a question that could create controversy in the context of business education (Andersson 2016). The following example involves a situation when such a controversy comes to the fore in embodied explorations of decision-making 'for sustainability'.

Business students and teachers of drama, business economics and civics were using a variation of Forum Play to explore situations in which business decisions were made (Textbox 19.2). The idea was to embody situations that the students

Textbox 19.2   Embodied explorations of decisions 'for sustainability' – step by step

1. Background knowledge and inspiration

Background knowledge about sustainability could be acquired through working with real or constructed cases, by reading media reports or by watching a documentary relating to sustainability issues (pesticides harming pollinating insects, the well-being of workers in large corporations, the well-being of animals or the protection of natural environments). A documentary about such issues that has been widely used in educational settings is the award-winning *The Corporation* (Achbar et al. 2003), which portrays the historical evolution of the large corporations we have today (including Dow Chemical mentioned in the introduction to this chapter).

2. Preparation of play

Students prepare a short play (5 minutes) that has an 'un-sustainable' ending. Situations from documentaries or case studies could give inspiration to particular situations and strategies to explore in more depth. In order to guide students to practice situations that could be particularly relevant in their future, the students could be 'given' specific roles to be included. For example, in the context of business education, roles like sub-contractor, business manager, employee, etc. could be 'provided'. Preparation of the play includes students pondering their roles: Who am I? Where am I? What do I want and why? What could make me change my behaviour?

3. Playing the play

A stage is created in the classroom with surrounding space making it easy for students (in Boal's terms spect-actors) to see and step up on the stage. The play is played out once, without interruption, in the classroom with classmates as audience. The role of the teacher here is to be facilitator, referred to by Boal as 'the joker'. This means taking responsibility for the process and helping the spect-actors solve the problem in plausible ways, explaining the rules and ensuring that all suggestions are respected and that everyone is given the opportunity to have a say. After the play is over, the teacher/joker asks questions to the students/spect-actors to clarify what happened in the play. What happened? What was 'unsustainable'? What would be (more) sustainable? Which role could or should try a different strategy?

4. Re-play of the play

The play is re-played with the difference that now the spect-actors can say 'stop' when something unsustainable happens or you or someone else finds that someone could act differently. The spect-actors can then either suggest different actions for the actors to carry out on stage in an attempt to change the outcome to something more sustainable, or come on stage to exchange the actor and perform their own interventions. This step is then repeated with the aim to explore as many strategies as possible. When making an intervention, the student/spect-actor can choose from where to start and is then allowed to try the new strategy without interruption from the spect-actors. After the intervention, the teacher/joker asks the actors involved in the 'unsustainable situation' about what they felt about the new strategy and thanks the spect-actor contributing with a strategy. If no spect-actor says 'stop' or the spect-actors find it hard to suggest a strategy, the joker (or a spect-actor) can say 'freeze' to give time to think and reflect about the situation, individually or in pairs, before going on. This step aims at finding as many strategies as possible, and when there are no more suggestions, it is time to step out of the roles.

5. More or less sustainable? analysing the played-out strategies

Here the teacher (who has stepped out of the joker role) leads the process of discussing and analysing or assessing the played-out strategies (involving different decisions) in terms of more or less sustainable. Apart from students' personal feelings in relation to sustainability, the human rights declaration, the child convention and/or sustainable development goals could be used to specify sustainability criteria that could be useful when assessing the strategies.

6. Ethical reflections of played-out strategies

For further reflection of the strategies played out in the Forum Play, the suggestions and situations where the spect-actors said 'stop' could be further analysed and reflected upon. What were the guiding principles or ethics behind the suggested strategies? For whom and when was the suggestion more or less sustainable? The principles for teaching ethics and morals presented in chapter 7 could be useful here.

found unsustainable and to explore how the situation could be made more sustainable and by whom. The students had prepared a play about environmental damage and young children working in dangerous conditions (Step 2, Textbox 19.2). They had played the play once (Step 3), re-played the play (Step 4) and a 'spect-actor' had said 'freeze' in a moment where the student in the role of executive manager of a factory producing garments for the corporation H&M was situated in a difficult position. The executive manager's difficult position related to conflicting expectations from shareholders, workers, NGOs and the sustainability manager. The workers' representative had convincingly described the dangerous working conditions; the representative from the environmental organisation had pointed to how the production process damaged the water source that the entire community was dependent on for their livelihood. In the 'freezed' situation, the teacher/joker asked the student in the role of executive manager about what it *felt* like, when the student objected in a trembling voice:

Student 2 (in role as executive manager): 'You're not allowed to ask that question . . . that's what it must be like'.[3]

How can we understand Student 2's utterance? Taking the broader context into consideration, we need to consider that the exercise was used in the context of a lesson in business economics and that it is reasonable to assume that the student was familiar with 'the business of a business is to do business' discourse. In this context, the student's objection can be understood in relation to an a-moral or a-political business ideal that does not 'allow' an executive manager to involve personal feelings when making business decisions. Accordingly, answering a question about one's own feelings was not possible in this situation, and from this perspective a teacher should not even be allowed to ask such questions. Accordingly, the student's reaction to the teacher's question could be described as contesting the idea of an executive manager in a business as a moral or political subject.

Business as un-usual and subjectification in classroom practice – a didactic model

In the two previous sections, we have met two students taking part in classroom practices where different worldviews come to the fore in conflicting standpoints as to whether trade automatically leads to socially and environmentally optimal outcomes or not. Student 1 expressed being confused when it was proposed that trade was not necessarily desirable (sustainable) in itself and Student 2 objected to a question about personal feelings when in role as an executive manager. The different and conflictual aspects that come to the fore in the first case relates to whether trade automatically leads to socially and environmentally optimal outcomes. In the second case, the different and conflictual aspects lie in whether an executive manager should involve personal feelings when making business decisions and act as a moral/political subject. Or, in other words, whether an executive manager should be guided by an a-moral/a-political *or* moral/political business ideal.

Subjectification as educational quality also comes to the fore in the moments highlighted in the practical examples. Drawing on the work of Laclau (1990), these moments could be described as dislocatory moments, i.e. situations when it is not clear how to 'go on' or engage in routine practices because a guiding principle (logic) is 'lost'. In the first case, it is the *loss* of a market liberal logic, and in the second case it is the 'apolitical ideal' that is at stake.

Laclau talks about moments of dislocation as 'traumatic' at the same time as some freedom is won. They are 'traumatic' in the sense that they deprive a person from an affective investment in a certain belief of how things are or ought to be. 'Freedom' is won because it offers some independence in relation to a social structure (socialisation). The loss can be described in terms of 'dis-identification' in relation to a logic in which one is affectively invested in, and to 'go on' it is necessary to 'identify anew' by affectively investing in another logic. Affective investment implies involvement of feelings and reason (body and mind) and is connected to subjectivity (Howarth 2013). A process of 'dis-identifying and identifying anew', which is called for by a dislocatory moment, is here understood as a subjectification process in which affective investment is facilitated and necessary. Accordingly, dislocatory moments are important in that they offer scope for students' subjectification processes at the same time as they can contribute to business as un-usual by opening up for alternative perspectives regarding how to organise social life. Accordingly, by creating conditions for and taking care of these moments, a teacher can open up for 'business as un-usual' without compromising emancipatory educational ideals. To facilitate this, I will here suggest a didactic model that can help teachers to recognise and take care of these moments. The procedure builds on a previously published research methodology (Andersson 2018). The model comprises three steps involving recognising (1) the emergence of a dislocatory moment, (2) its closure and (3) the change of logics facilitated by a dislocatory moment. I will use the examples from the previous two sections to illustrate the procedure.

Step 1: emergence of a dislocatory moment

This step involves identifying the guiding principles (i.e. logics) that the students are invested in and identifying the emergence of a dislocatory moment. This is done by paying attention to and analysing students' ways of talking about what they should do or what others in a specific role or position should do, explicitly or implicitly. This could involve, for instance, paying attention to ways of talking about the role of business in relation to sustainability. Are students expressing uncertainty, frustration or self-confidence? Are there any signs of 'voids' that occur when students' guiding principles (i.e. logics or worldviews) are lost? Are there any expressions that in Laclau's terms could be described as signs of minor 'traumatic' (or troublesome) experiences? Being attentive to signs of uncertainty, frustration, decrease of self-confidence, 'voids' or what could be described as expressions of 'minor traumatic experiences' also requires listening carefully to the tone of the voice and the body language. This is possible for teachers (or an observer) to do but hard or impossible just from reading a transcript. With this in mind, Student 1's literal expression of *confusion* and *hesitation* when participating in the four-corner exercise (see '"Four corners" – to experience the political dimension of sustainability issues') and Student 2's *trembling* voice when invited to describe personal feelings when playing the role as executive manager (see 'Embodied explorations of decision-making') are here interpreted as signs of emerging dislocatory moments. Identification of a dislocatory moment 'in full' also requires analysis of the guiding principles (i.e. logics) in students' reasoning before and after these moments. This is done in the next step analysing guiding principles in Student 1's reasoning as an illustrative example.

Step 2: closure of a dislocatory moment

The second step involves analysis of how the dislocatory moment is closed by students investing in or identifying with (other) logics. How do students fill the void that the dislocatory moment makes visible? What facilitates the closure of a dislocatory moment? Which new logics can be discerned? This analysis involves listening to the students taking part in different oral lesson activities and reading written assignments paying attention to when they invest in another guiding principle. By listening to Student 1 taking part in classroom activities and reading Student 1's written assignments, a closure of the dislocatory moment could be identified in a lesson when the class was working on a real case where many textile workers died in a factory fire. The student here suggested that and how the business management should 'handle the situation' by taking better control of the supply chain. The student's suggestion, expressed in a tone of confidence, is here interpreted as a sign of closure.

Step 3: change of guiding principles/logics

The last step involves analysis of the change in the guiding principles that the student identifies with prior to and after the dislocatory moment. This involves listening to the students taking part in different oral lesson activities and reading written assignments while focusing on the guiding principles underpinning the students reasoning. By listening to and reading Student 1's work in this way, the dislocatory moment could be described as the beginning of this student's process to acquire a variety of perspectives regarding the role of business[4] in relation to sustainable development. The role of a businessperson that was facilitated through the dislocatory moment could be described as one who, as owner, can be a powerful change agent that can and should take responsibility by taking control of the supply chain, contribute to the common good and consider fair distribution of profits.

The student's process that started with a dislocatory moment could be described in terms of a subjectification process, which involves dis-identification and identifying anew by affectively investing in another guiding principle (logic). The aspects of dis-identification in the above examples involves dis-identification with a market liberal logic, and dis-identification 'from the norm the teacher instilled' (see also chapter 4 and Håkansson et al. 2018, 106). Identifying anew in the practical example involves identifying with a logic positioning a businessperson as a powerful change agent driven by a purpose of contributing to the common good. The scope for subjectification is connected to the dislocatory moment, which means that subjectification 'occurs' between the emergence (step 1) and the closure (step 2). When aiming to offer scope for students' subjectification processes, it is therefore important for the teacher to take some time to 'stand by' and give students time to stay in the moment, without leaving them completely. This requires sensitivity from the teacher and is not an easy task. My hope is therefore that this didactic model building on the concept of 'dislocatory moments' could be helpful for teachers in collegial conversations about approaching this difficult task.

Concluding comments

I have here suggested teaching approaches that could offer space for students' subjectification processes, opportunities to discover and clarify different and conflicting perspectives on sustainability problems, and embodied meaning-making (or learning understood in a broad sense) of 'sustainable' decisions. When using these methods in the classroom, the aim is to clarify and explore a variety of perspectives, values or different decisions that could be made in particular situations. In this sense, there are no right or wrong solutions when working with these methods. However, this does not mean that all positions or suggested strategies are equally valid in relation to sustainability problems (neither is a student's change of 'guiding principle' automatically a change for sustainability). To avoid an 'anything-goes' approach, it is therefore important to consider when to use these exercises and what to do next. In the variation of Forum Play presented in this chapter, I have therefore included a step that involves analysis or assessment in terms of more or less sustainable, and suggested further ethical reflection in line with the principles described in chapter 7.

Acknowledgements

I wish to thank teachers and students for sharing their work in classroom practice. I also wish to thank colleagues at my department for constructive comments from the perspectives of being educators, PhD students and researchers in drama and educational sciences.

Notes

1 The word 'clarify' could give the impression that values are stable essential entities only to be discovered; however, that is not the intention. Values are here seen as changeable/contingent and they can for instance change when listening to other peoples' perspectives during a four-corner exercise.
2 The reader familiar with the writings of Boal could object to using Forum Theatre and in the way that is was used in the example since the actor addressed as 'oppressed' is the 'business manager', and that it in line with Boal could be described as fundamentally at odds to think of a business person in terms of oppressed. However, there is a similarity in that one could describe this as a kind of 'discursive oppression' where 'the oppressor' could be described as 'the a-political ideal´ (Andersson 2016) which obstructs a business's actions as moral or political subjects.
3 The quote has previously been used in Andersson (2016, 15).
4 A business role is here understood as a subject position which is further described in chapter 15.

References

Achbar, M., Abbott, J. and Bakan, J. (2003) *The Corporation*. Zeitgeist Films, New York.
Andersson, P. (2016) *The Responsible Business Person: Studies of Business Education for Sustainability*. PhD Thesis, Södertörn University, 116.

Andersson, P. (2018) "Business as un-usual through dislocatory moments: Change for sustainability and scope for subjectivity in classroom practice". *Environmental Education Research*, 24(5): 648–662.

Beck, U. (2009) *World at Risk*. Polity Press, Cambridge.

Biesta, G. (2014) *The Beautiful Risk of Education*. Paradigm Publishers, Boulder.

Boal, A. (2008) *Theatre of the Oppressed*. Pluto Press, London.

Byréus, K. (2001) *Du har huvudrollen i ditt liv: Om forumspel som pedagogisk metod för frigörelse och förändring* [You have the leading part in your life: About forumplay as pedagogical method for emancipation and change]. Liber, Stockholm.

Dryzek, J.S. (2013) *The Politics of the Earth: Environmental Discourses*. Oxford University Press, Oxford.

Glynos, J. and Howarth, D. (2007) *Logics of Critical Explanation in Social and Political Theory*. Routledge, London.

Håkansson, M., Östman, L. and Van Poeck, K. (2018) "The political tendency in environmental and sustainability education". *European Educational Research Journal*, 17(1): 91–111.

Howarth, D.R. (2013) *Poststructuralism and After: Structure, Subjectivity and Power*. Palgrave Macmillan, New York.

Kopnina, H. (2012) "Education for Sustainable Development (ESD): The turn away from 'environment' in environmental education?" *Environmental Education Research*, 18(5): 699–717.

Laclau, E. (1990) *New Reflections on the Revolution of Our Time*. Verso, London.

Österlind, E. (2011) "Forum play: A Swedish mixture for consciousness and change". In Schonmann, S. ed., *Key Concepts in Theatre/Drama Education*. Sense Publishers, Rotterdam, 247–251.

Porter, M.E. and Kramer, M.R. (2011) "Creating shared value". *Harvard Business Review*, 89(1–2): 62–77.

Rundlof, M., Andersson, G.K.S., Bommarco, R., Fries, I., Hederstrom, V., Herbertsson, L., Jonsson, O. et al. (2015) "Seed coating with a neonicotinoid insecticide negatively affects wild bees". *Nature*, 521(7550): 77–80.

Sund, L. and Öhman, J. (2014) "On the need to repoliticise environmental and sustainability education: Rethinking the postpolitical consensus". *Environmental Education Research*, 20(5): 639–659.

Wals, A.E.J. (2010) "Between knowing what is right and knowing that it is wrong to tell others what is right: On relativism, uncertainty and democracy in environmental and sustainability education". *Environmental Education Research*, 16(1): 143–151.

20 Political emotions in environmental and sustainability education

Ásgeir Tryggvason and Andreas Mårdh

Introduction

When the political dimension of environmental and sustainability education (ESE) is acknowledged, the question of what role emotions should play in the classroom becomes pressing (Håkansson and Östman 2018). As we have seen in previous chapters (chapters 8 and 13), the political dimension of sustainability issues can be tightly intertwined with students' emotions toward these issues. We know, for instance, that students' emotions toward sustainability issues are important for how their engagement (or non-engagement) in sustainability issues takes shape (Ojala 2012). In classroom discussions, heated emotions can arise that require the teacher to approach emotions in an educational way. But classroom discussions can also be characterised by the students' lack of engagement and emotional involvement, which requires other kinds of approaches. As we see it, a main question is then how teachers can approach emotions in ESE in a non-instrumental way and at the same time recognise their educational potentialities.

In this chapter, we focus particularly on the political dimension of students' emotions in terms of 'political emotions' in ESE practice. The purpose is to provide the reader with two strategies for approaching 'political emotions' in ESE practice. The strategies can be seen as suggestions of how to approach and educationally elaborate with political emotions in the classroom.

The main concept in this chapter is 'political emotions'. The concept itself is however debatable, and several scholars use different, and sometimes conflicting, definitions (see Håkansson and Östman 2018; Mouffe 2014; Ruitenberg 2009; Tryggvason 2017). Without going into a conceptual debate, we will use two criteria to define political emotions. *Political emotions* are those emotions that

(a) revolve around the boundaries between 'us' and 'them' and
(b) relate to substantially different visions of society.

Criteria (a) means that political emotions revolve around some form of 'us' and 'them' formations. In this way, political emotions are by definition bound up with the formation of collective identities. Criteria (b) stipulates that the boundary between 'us' and 'them' must stem from different visions of society. This means that the difference between 'us' and 'them' ultimately stems from the difference between what kind of society 'we' want and what kind of society 'they' want (see Ruitenberg 2009; Tryggvason 2017; Van Poeck and Östman 2018). An *emotion* is here defined as a bodily experience that a person is aware of; in other words, an experience that a person feels. Emotions are in this sense not just reactions or unconscious affects (see chapter 10), but are experiences that a person knows something about (even if we don't always know why we feel as we do). Furthermore, emotions are directed toward something or someone (in contrast to a mood, which is not directed toward something particular). This definition gives a rather common sense idea of what an emotion is; if you are *aware* that you are angry about *something*, then this bodily experience could be defined as an emotion.

The two strategies that we outline in this chapter stem from research on political emotions in teaching practices (Håkansson et al. 2018; Sund and Öhman 2014). The first strategy is *simplification*, which highlights the relation between complex sustainability issues and students' emotions and involvement in sustainability issues (Mårdh and Tryggvason 2017). The second strategy is *circulation*, which highlights how emotions can circulate in the classroom by being attached to objects that carry an emotional history (Ahmed 2014).

As teachers we find that political emotions, despite their potential risks, should not be suppressed or downplayed, but instead seen as productive elements that already are present in the ESE classroom. However, by seeing political emotions as productive elements does not mean that these emotions always should run high in the classroom. Instead, the two strategies that we suggest are about enabling teachers to handle and approach political emotions in their classroom. This means that the strategies are aimed at enabling us to make thoughtful and considered choices about when and how to approach political emotions in the classroom. In the following section, we outline the two strategies as a way to approach and reflect upon political emotions in ESE practice.

Teaching with and through political emotions in ESE

By outlining the strategies as ways to teach *with* and *through* political emotions, we want to avoid a purely instrumental perspective on emotions in education. In line with Ojala (2017) we do not see emotions just as means for some educational goals, such as learning more about sustainability issues. Instead, we see political emotions as being vital to a vibrant democratic life as well as to an engaging ESE practice. In order to have strategies on how to teach with and through political emotions we start by outlining *simplification* and thereafter *circulation*.

Simplification as a strategy for mobilising students' political emotions

In outlining *simplification* as a strategy for reflecting on how to mobilise students' political emotions in ESE, we will take our starting point in a classroom scenario.

Imagine a situation where emotions of every kind seem to be absent in the classroom. In this situation, the teacher brings up what s/he thinks is a burning environmental issue, but the students seem somehow emotionally indifferent. The students discuss the issue with each other and they present different perspectives on the topic, as well as thoughtful insights to the discussion; in that sense they are definitely learning *about* the environmental issues. But as the students analyse and describe how the issues have evoked heated political conflicts in the society, they don't seem to get involved in any conflicts over the issue themselves. In other words, the students seem to analyse the society from a distance, as if they themselves were not a part of it. When the lesson ends, the teacher knows that the discussion will end as soon as the students leave the classroom.

As a teacher in this scenario, we would probably want the students to get more emotionally involved in the issues they are discussing and would like to enable them to picture themselves as being part of the society and the environment. A main question is therefore how to get the students more engaged and emotionally involved. To elaborate on this question, we will direct our attention toward the issue itself and its characteristics, rather than toward the students and their lack of emotional involvement.

In ESE practices, a pluralistic tradition can be seen as one of the more common approaches in teaching sustainability (see chapter 5). From a pluralistic perspective, sustainability issues are seen as complex and conflictual, and they don't come with simple solutions. Moreover, sustainability issues are seen as issues where economic, social and environmental aspects of a problem are closely intertwined and where it is difficult to approach the issue from merely one of these vantage points. As sustainability issues often involve several conflictual aspects, they often run into an ambiguity in terms of 'on the one hand . . . but on the other . . .'.

To illustrate this complexity in a sustainability issue, a concrete example could clarify. Imagine a teaching situation that is about the explosion of the oil rig Deepwater Horizon and the following oil spill in the Gulf of Mexico in 2010. This case includes myriad economic, social and environmental aspects. There are multiple perspectives and ways to introduce this as a teaching topic in a classroom, and each aspect that is emphasised tends to downplay other aspects. For instance, how should it be described and worded when introduced to the students: As an accident? As a disaster? Or as an environmental crime? From a pluralistic perspective, the teacher would highlight exactly this complexity and the multiple aspects that an issue can have. Thus, the teacher would not aim to reduce the dilemmas, paradoxes and complexities that the case brings but instead place them in the foreground.

But emphasising the complexities and ambiguities in an issue such as the Deepwater Horizon case could be exactly the act that prevents students from being moved by the issue itself. Perhaps it is not so surprising that students don't get emotionally involved in an issue when they have no safe side to stand on, and where there always is an 'on the other hand . . .' (see Mårdh and Tryggvason 2017, 611). It is in relation to this teaching problem that we put forward *simplification* as a strategy for reflection on how to teach with and through political emotions.

We are aware that the term 'simplification' can signal that this is about treating complex issues in a non-reflective, or even banal, manner. However, this common sense use of the word is not what we are after. When we use the term 'simplification' in this chapter, it is as a specific theoretical concept that stems from political philosophy, which we here formulate as a strategy for teaching. With this said, simplification can be characterised as consisting of two different political moves (see chapter 13). The first kind of moves is when the teacher *draws a line* in order to simplify the conflictual aspect of the specific teaching content. This means that the teacher highlights that a conflictual line can be drawn between those collective identities involved in the issue or problem. Here it is important to underscore that such a line is not about identifying essential differences between collective identities; on the contrary, simplification is about offering political positions that have no other ground than the very act of taking a stand. In this way, the positions that this 'drawing of a line' enables are political positions. As political positions, they do not depend on who the others *are*, but are instead formulated around the difference between what 'we' and 'they' *want* (Mouffe 2005; Todd 2010).

The second kind of moves is when the teacher equalises differences in order to simplify the complexity of the teaching content. This means that the complexity of a sustainability issue is arranged in such a way so that all the different parts that constitute this complexity are arranged as being equivalent to each other. This may seem a bit vague, but what we want to point to here is that simplification should not be understood as an approach that reduces the differences between perspectives on an issue. Instead, simplification is about arranging differences as being in an equivalent relation to each other (Laclau and Mouffe 1985/2001; see also Mårdh and Tryggvason 2017). Every sustainability issue can be seen from multiple perspectives. By placing different perspectives as being equivalent to each other means that they are arranged as being of the same kind, even if there is a difference between them. Thus, the difference is not reduced, but is arranged in a certain way that simplifies the complexity of the issue.

In brief, *simplification* offers two kinds of moves; moves where the teacher *draws a line* in order to simplify the conflictual aspects of an issue, and moves where the teacher equalises differences in order to simplify the complexity of an issue. In order to exemplify what these two kinds of moves mean, we will continue with the Deepwater Horizon case.

Simplifying the conflictual aspects of this case by *drawing a line* could, for instance, be the line between the categories 'people' and 'profit' (see chapter 13). Drilling up oil from the bottom of the sea comes with the calculable risk that the pursuit of profit can harm people's livelihood and environment. From this perspective, to either be for or against the drilling of oil cannot merely be grounded in facts, but must ultimately rest on the very act of taking a stand. This means that the political dimension of the issue rests upon nothing else than the act of making a decision in a terrain of un-decidability.[1] So instead of presenting the positions 'people' and 'profit' in terms of 'on the one hand . . .' and 'on the other hand . . .', the drawing of a line articulates this conflict as an *either-or conflict* (cf. chapter 13). Underscoring the conflictual aspects of environmental and sustainability issues could be seen as particularly important, as conflicts tend to be downplayed in students' discussions of sustainability issues (Öhman and Öhman 2013) as well as in educational policy (Huckle and Wals 2015). It is also crucial that we keep in mind that the very concept 'sustainability' is tightly linked to the idea of consensus rather that conflict (see chapter 1).

Simplifying the complexity of the case with moves that *equalise differences* can be exemplified in how the case is related to other cases. The case of Deepwater Horizon can be listed as one of many cases of oil spill that have affected environment and societies. Equalising moves places this case alongside other 'similar' cases. Here, the particular complexity that characterise the Deepwater Horizon case is not reduced but arranged as being equivalent to the complexities of other cases. This means that the complexity of the case is maintained while arranged as being equivalent to other cases. A key aspect is the teacher's decision of what to place as an equivalent case, and the way the teacher initially presents the Deepwater Horizon case determines which other cases can be seen as equivalent to it. For example, if the teacher decides to present the Deepwater Horizon case primarily as an environmental accident, then equivalent cases consist of other environmental accidents. If the Deepwater Horizon case instead is presented primarily as an environmental crime, then other cases can be seen as equivalent to it. However, the decision of how to present a case, and the decision of what other cases are equivalent to it, is ultimately a decision that comes with a political dimension that cannot be evaded. These decisions, and their political dimension, can be seen as an integral part of teachers' professional practice in ESE. The strategy of *simplification* is one way to reflect on these decisions. Another strategy is *circulation*, to which we now turn.

Circulation as a strategy for maintaining students' political emotions

In the previous section, we outlined simplification as a strategy for mobilising political emotions and engaging students in environmental and sustainability issues. In this section, we will focus on situations where political emotions are *already* in play and where we as teachers instead want to maintain and (re)orientate them for educational purposes. After all, it is not uncommon for students to collectively express political emotions and, consequently, be at odds with either each other or some external adversary putting their vision of a sustainable society into question. Imagine, for instance, a classroom scenario where the students discuss topics such as meat consumption, the increasing use of fossil fuel or other issues that are not only burning environmental and sustainability issues but are also issues that relate to the students' own habits of living. In this classroom scenario, the teacher notices that the students no longer approach sustainability with analytical distance but that they instead are passionately and emotionally invested in taking political stances. When faced with this situation, the teacher may want to maintain the students' level of seriousness and involvement as their discussion takes the shape of a strong and genuine educational experience. At this point, a pressing question for the teacher is how s/he can facilitate and maintain these political emotions.

As in the previous section, we will here put forward a concept which we have found particularly helpful in reflecting on scenarios of the sort depicted above. The concept we suggest is *circulation*, as it has been introduced by cultural theorist Sara Ahmed (2014). With this concept, we argue that it is possible to reflect on how to *maintain* and *orientate* political emotions in the classroom, even when it comes to difficult ones such as hope and fear.

To begin with, emotions are from this perspective not seen as the property of any one individual or as something that can inherently reside in things. Instead, they are *social* in the sense that they take shape through the continuous encounters between subjects, objects and the narratives that bring them into contact with each other. However, emotions are not only shaped by these encounters but are themselves constitutive, meaning that objects and individuals become *specific* objects and individuals through the very relationships that they share. In this sense, the 'I' and 'we', as well as the 'you' and 'them', is constantly being made and remade by the way in which our interactions are carried out (Ahmed 2014).

From this follows that emotions are always directed towards objects or other beings in the sense that we feel a particular emotion for some*thing* or some*one*. This raises the question about how we come to attach our emotions to certain things but not others. The answer to this question is that every such attachment is the result of a historical process where an object of emotion has continuously circulated over time, between contexts and through the repetition of associative narratives so that it gradually has become saturated with affect. In other words, the object has acquired a heightened ability to move us into action (Ahmed 2014).

One way to think metaphorically about emotions is to imagine them as part of an economy where objects gradually accumulate affective value the more they circulate in our society. Or, as Ahmed (2014) puts it: 'emotions work as a form of capital: affect does not reside positively in the sign or the commodity, but is produced as an effect of its circulation' (p. 45). In short, circulation creates a sort of surplus value where objects become more and more affective the more they are coupled with a given set of narratives.

In close relation to this affective economy is the so-called fetishisation of objects of emotion. The concept of fetishisation points to how we take for granted what we 'naturally' feel for an object or for another being. In this sense, the concept highlights our acts of forgetting the process through which an object has become affective, for instance when we come to hate, fear or love something without considering how and why we have come to feel this way. Thus, fetishisation is the height of circulation where we are left with only the emotion while its history of production is overlooked and forgotten (Ahmed 2014).

As with any economy, the affective kind also has a political dimension in the sense that its effects are unevenly distributed. For instance, imagine a scenario where you already before encountering an object assume that it is to be feared;

you will most likely become orientated away from that thing without necessarily having experienced it first-hand. In some cases, this may be a good tendency (as in moving away from dangerous objects), but in other cases it becomes problematic. For example, when it comes to an issue such as deeply entrenched racism, where feelings of hate or disgust tend to orientate people away from each other as well as facilitate a social exclusion, it becomes clear how the uneven effects of circulation can be highly problematic (Ahmed 2014).

So far, our outline here has mostly been an introduction to the concept of circulation. Therefore, we will now turn to different examples of how circulation can be a strategy for ESE practices. Drawing on the account above, an educational use of circulation would first and foremost mean that we as educators try to identify the emotional flows that are already present in our classrooms, and from this either preserve or change their direction. In short, it entails searching for what kind of objects students already feel strongly about as well as considering the political consequences that follow from their emotional investments.

From here, teachers can basically take two courses of action, or as Katrien Van Poeck and Leif Östman (see chapter 13) label it in the context of ESE, the teacher can make certain *moves* that either narrow or broaden the space for political emotions relating to sustainability. Building on their concept, as well as the account given above, in the following paragraphs we will talk more about how teachers can make use of *confirming* and *historicising* moves in ESE practice.

First, if a teacher, using his or her professional judgement and knowledge of curricular aims, finds that the direction of students' emotions are suitable for sustainable living, then s/he would do well in trying to maintain the current circulation. This could be done by retelling narratives that have a history of being associated with the object of emotion.

To exemplify: say that a group of students participating in a classroom discussion are already heavily invested in trying to transform society toward more renewable energy sources. We could imagine, for instance, that their hope for the future is attached to solar technology while their fear and anger are bound up with nuclear power or fossil fuel consumption. Given this, the teacher can aim at increasing the affective value of these objects through *moves that confirm* and acknowledge the troublesome effects of unsustainable energy sources. This could, for example, be done by pointing to instances of nuclear power failures (such as Three Mile Island, Chernobyl and Fukushima) and place them in juxtaposition to the recent advancements made in solar power technology.

Secondly, if the teacher wants to maintain the intensity of students' political emotions but nonetheless change its *orientation*, it becomes his or her task to 'demystify' the object of emotion by showing its history of production. For instance, imagine a group of students that are emotionally invested in a dietary habit which the teacher, through greater experience, knows not to be sustainable. Perhaps the students feel such great love for meat consumption (as well as aversion towards those who try to reduce society's consumption of it) that meat as a product and foodstuff has become a fetish.

In this case, the teacher can attempt *moves that historicise* the object by highlighting how meat consumption has become 'naturally' associated with a specific set of emotions. This could mean that the teacher, in highlighting the process of affective accumulation, shows the students that they encounter the object through commercial narratives that associate meat eating with a happy life. Moreover, the teacher could point to how their emotions toward meat consumption are embedded in medical narratives that stresses the health benefits (especially for children) of consuming meat. As such, historicising moves require the teacher to demonstrate how the objects of emotion are not static by nature but rather produced over time and that, in this case, meat consumption has become associated with happiness, wealth and health through the continuous repetition of certain narratives. In short, the challenge and task of the teacher can be summarised as follows: 'Once an affective quality has come to reside in something, it is often assumed as without history. We need to give this residence a history' (Ahmed 2014, 214).

Thus, we believe that historicising moves can enable teachers to open up a space for students to consider new objects of emotion whose political consequences for sustainability are, if not less severe, then at least different in character. But because there are no guarantees that teachers fully and with precision can control the circulation of emotions, our main point is rather that these moves enable the students to experience a sudden un-decidability in a political terrain that nonetheless requires of them to act or to make a decision (Håkansson and Östman 2018).

Summary

In this chapter, we have outlined two strategies for teaching with and through political emotions in ESE practices. These two strategies are summarised below:

> *Simplification*: In a classroom situation where students seem disengaged and indifferent toward sustainability issues, simplification is a strategy that aims at mobilising students' political emotions by simplifying the conflicts and the complexity of the issue. This strategy comes with two kinds of moves.
>
> - *Moves that simplify the conflict by drawing a line.*
> - *Moves that simplify the complexity by equalising differences.*
>
> *Circulation*: In a classroom where students already are emotionally involved in the issue, circulation is a strategy for the teacher to maintain the intensity of the emotions and (re)orientate them toward other objects. This strategy also comes with the two kinds of moves.
>
> - *Moves that confirm the intensity of students' emotions.*
> - *Moves that historicise students' emotions and (re)orientate them toward other objects.*

To conclude, what we want to emphasise by outlining these two strategies is that it is crucial to acknowledge and identify the educational possibilities that emotions can bring in ESE practices. As educational suggestions, these strategies are open for teachers' own modification and revisions. We are well aware that the educational settings and teaching contexts can differ tremendously, and at the end of the day it is therefore always our professional judgement as teachers and our contextual awareness that is key in enacting these strategies when teaching with and through political emotions.

Note

1 We are of course aware of the crucial role that facts have in the act of taking a stand. Our point here is that the act cannot rest upon the facts themselves.

References

Ahmed, S. (2014) *The Cultural Politics of Emotion*. Edinburgh University Press, Edinburgh.

Håkansson, M. and Östman, L. (2018) "The political dimension in ESE: The construction of a political moment model for analyzing bodily anchored political emotions in teaching and learning of the political dimension". *Environmental Education Research*, pre-published online: https://doi.org/10.1080/13504622.2017.1422113

Håkansson, M., Östman, L. and Van Poeck, K. (2018) "The political tendency in environmental and sustainability education". *European Educational Research Journal*, 17(1): 91–111.

Huckle, J. and Wals, A.E.J. (2015) "The UN decade of education for sustainable development: Business as usual in the end". *Environmental Education Research*, 21(3): 491–505.

Laclau, E. and Mouffe, C. (1985/2001) *Hegemony and Socialist Strategy: Towards a Radical Democratic Politics*. Verso, London.

Mårdh, A. and Tryggvason, Á. (2017) "Democratic education in the mode of populism". *Studies in Philosophy and Education*, 36(6): 601–613.

Mouffe, C. (2005) *On the Political*. Routledge, New York.

Mouffe, C. (2014) "By way of a postscript". *Parallax*, 20(2): 149–157.

Öhman, J. and Öhman, M. (2013) "Participatory approach in practice: An analysis of student discussions about climate change". *Environmental Education Research*, 19(3): 324–341.

Ojala, M. (2012) "Hope and climate change: The importance of hope for environmental engagement among young people". *Environmental Education Research*, 18(5): 625–642.

Ojala, M. (2017) "Hope and anticipation in education for a sustainable future". *Futures*, 94: 76–84.

Ruitenberg, C.W. (2009) "Educating political adversaries: Chantal Mouffe and radical democratic citizenship education". *Studies in Philosophy and Education*, 28(3): 269–281.

Sund, L. and Öhman, J. (2014) "On the need to repoliticize environmental and sustainability education: Rethinking the postpolitical consensus". *Environmental Education Research*, 20(5): 639–659.

Todd, S. (2010) "Living in a dissonant world: Toward an agonistic cosmopolitics for education". *Studies in Philosophy and Education*, 29(2): 213–228.

Tryggvason, Á. (2017) "The political as presence: On agonism in citizenship education". *Philosophical Inquiry in Education*, 24(3): 252–265.

Van Poeck, K. and Östman, L. (2018) "Creating space for 'the political' in environmental and sustainability education practice: A political move analysis of educators' actions". *Environmental Education Research*, 24(9): 1406–1423.

Index